Supply Chain Management

Beiträge zu Beschaffung und Logistik

Reihe herausgegeben von

Michael Eßig, München, Universität der Bundeswehr, Neubiberg, Deutschland

Wolfgang Kersten, Institut Logistik & Unternehmensführung, Technische Universität Hamburg, Hamburg, Deutschland

Industrielle Wertschöpfung wird immer komplexer. Der steigende Wettbewerbsdruck zwingt zu differenzierten Angeboten, gleichzeitig nimmt der Kostendruck zu. Unternehmen können diesen gestiegenen Anforderungen nur gerecht werden, wenn sie neben der Optimierung eigener Produktion besonderen Wert auf die Gestaltung effektiver und effizienter Netzwerke legen. Supply Chain Management befasst sich mit unternehmensübergreifenden Wertschöpfungsaktivitäten von der Rohstoffgewinnung bis zur Endkundendistribution. Die Schriftenreihe sieht sich dabei besonders den lange vernachlässigten betriebswirtschaftlichen Teildisziplinen Beschaffung und Logistik verpflichtet, die als Treiber des Supply Chain Management gelten.

Ludwig Häberle

Truck-Sharing in Road Freight Transport

Springer Gabler

Ludwig Häberle
School of Management
Universität St. Gallen
St. Gallen, Switzerland

Dissertation Universität St. Gallen 2025

ISSN 2627-292X ISSN 2627-2938 (electronic)
Supply Chain Management
ISBN 978-3-658-48753-9 ISBN 978-3-658-48754-6 (eBook)
https://doi.org/10.1007/978-3-658-48754-6

This Springer Gabler imprint is published by the registered company Springer Fachmedien Wiesbaden GmbH, part of Springer Nature.
The registered company address is: Abraham-Lincoln-Str. 46, 65189 Wiesbaden, Germany

Vorwort

Sharing of trucks is caring about efficiency

Diese Dissertation ist das Resultat meiner gut dreieinhalbjährigen Forschungstätigkeit an der Universität St.Gallen. Während dieser Zeit durfte ich am Institut für Supply Chain Management (ISCM-HSG), getreu dem Motto "science based practice driven", eine Reihe von Forschungs- und Praxisprojekten begleiten, die mich fachlich und persönlich geprägt haben. Gerne möchte ich mit Blick auf diesen für mich prägenden Lebensabschnitt die Gelegenheit nutzen, einigen Weggefährten und Unterstützern meinen persönlichen Dank auszusprechen.

Allen voran gilt mein besonderer Dank meinen beiden Referenten, Andreas Herrmann und Torsten Tomczak. Nicht zuletzt für die Übernahme des Betreuungsverhältnisses in einer für die gesamte Universität aussergewöhnlichen Lage, welche durch die unmittelbare Betroffenheit kurzzeitig auch eine persönliche Unsicherheit für die vorliegende Dissertation darstellte. Ich möchte mich für das entgegengebrachte Vertrauen, die Unterstützung auf diesem Weg und den wertschätzenden Umgang bedanken. Ich habe es sehr geschätzt und als einmalig erlebt, am Institut für Mobilität (IMO-HSG) eine zweite institutionelle Heimat zu finden. In diesem Zuge gilt mein Dank auch Philipp Scharfenberger, für den kreativen Austausch und das inhaltliche Sparring zur Konzeption dieser Dissertation.

Des Weiteren danke ich Thomas Friedli, welcher die Leitung des ISCM-HSG in einer herausfordernden Zeit übernommen und ein Umfeld der Stabilität geschaffen hat. Durch sein persönliches Engagement hat er uns Mitarbeitenden am Institut Struktur und Sicherheit vermittelt, was eine Grundvoraussetzung für den Erfolg einer jeden Dissertation darstellt.

Ein besonderer Dank gilt Wolfgang Stölzle, dessen Wirken und persönlicher Einsatz von grundlegender Bedeutung für den Erfolg dieser Arbeit ist. Als langjähriger Begleiter während meiner Zeit am ISCM-HSG, seit dem Eintritt als Werkstudent parallel zum Masterstudium im Dezember 2018, habe ich stets die vertrauensvolle und intensive Zusammenarbeit geschätzt. Als Impulsgeber für die Dissertation und Mentor war er in fachlicher und persönlicher Hinsicht prägend für meine persönliche Weiterentwicklung.

Aufgewachsen mit und geprägt vom familiären Speditionsgeschäft war und ist mir der Praxisbezug ein persönliches Anliegen, insbesondere auch zu den zahlreichen kleinen und mittelständischen Unternehmen, welche die Transportbranche prägen. Insofern gilt mein Dank auch der Vielzahl an Praxisvertretern aus Transport und Logistik, Industrie, Handel sowie von Verbandsseite, für den intensiven, konstruktiven, kritisch hinterfragenden und offenen Austausch in Interviews, Workshops und darüber hinaus.

Was wäre eine Promotion ohne das persönliche Umfeld aus Kollegen, Freunden und Familie? Zuallererst möchte ich ein herzliches Dankeschön an meine ISCM-Institutskollegen Tim Auer, Calvin Klein, Petra Lehmann, Leon Zacharias und Laurin Zemmrich aussprechen. Die gemeinsame Zeit und vor allem die gemeinsamen Erlebnisse werden mir stets in guter Erinnerung bleiben. Des Weiteren danke ich den "Ehemaligen", Joschka Beck, Maximilian Enthoven, Daniel Langner, Raphael Preindl und Stefan Selensky. Nicht zuletzt danke ich meinen Kollegen an meiner zweiten Institutsheimat, dem IMO-HSG. Insbesondere gilt mein Dank Andreas Collet und Marvin Greifenstein für die Anregungen aus etlichen Kreativphasen und Diskussionen. Zudem möchte ich mich bei Yannik Kohleisen, Julian Schließmann, Jonas Gottsche und Valentin Züsli bedanken, die als Praktikanten und Werkstudenten eine wertvolle Stütze in der Projektarbeit waren.

Von Herzen gilt mein Dank Dir, Annabel. Du hast mich auf diesem Weg begleitet und tagtäglich unterstützt, wann, wo und wie auch immer es nötig war. Du hast massgeblichen Anteil an dieser Arbeit. An dieser Stelle möchte ich auch den Wiegands meinen Dank aussprechen.

Abschliessend gilt mein Dank meinen Grosseltern, meiner Grosstante, meinen Eltern und meiner Schwester. Familiärer Rückhalt ist von unschätzbarem Wert und war mir immer eine grosse Stütze in allen Lebenslagen. In Höhen wie in Tiefen und weit darüber hinaus habt ihr mich stets bedingungslos unterstützt, ermutigt, meinen eigenen Weg zu gehen und damit all dies erst ermöglicht. In tiefer Dankbarkeit ist die Arbeit Euch allen gewidmet.

Zürich
im August 2024

Ludwig Häberle

Abstract

Europe's road freight transport industry is facing challenges such as an increasing shortage of truck drivers, low profit margins and growing freight volumes in light of infrastructural constraints. The ongoing shift towards zero-emission vehicles poses a further long-term driver of change. Although the optimisation of truck capacity utilisation has the potential to yield economic benefits for fleet operators and ecological improvements for the transport sector, the issues of empty running and load space utilisation efficiency persist. Despite growing pressure on efficient truck deployment, shared mobility concepts are still scarce in road freight transport—contrary to the increasing adoption of the sharing economy in other fields.

Addressing existing and emerging challenges in road freight transport, this thesis investigates the potential of truck-sharing as a strategy to enhance economic efficiency in truck fleet utilisation. The objective of this research is to develop a forward-thinking conceptualisation of truck-sharing that addresses inefficiencies in freight transport operations. This thesis is grounded in entrepreneurial practices building on the established concepts of horizontal cooperation and the sharing economy and employs a design-oriented approach. The research scope comprises three areas of investigation: (1) the exploration of application potential and future use cases, (2) the investigation of drivers and barriers, and (3) the derivation of implications and action-oriented recommendations. To this end, the research process follows an exploratory, mixed-methods design, integrating the engaged scholarship approach with backcasting methodology. The applied research methods include case studies with truck fleet operators, expert interviews, and an online survey of logistics service providers and truck drivers.

The result of this work is a conceptualisation of truck sharing that is oriented towards the future. When implemented, it has the potential to contribute to a reduction in the downtimes of freight vehicles and an increase in operational flexibility. Based on the current situation, this study identifies functional and perceptual barriers that require attention from multiple perspectives and resolution through measures at the organisational and individual levels. By proposing solution approaches, this dissertation establishes the foundation for the sharing of freight vehicles between transport companies as a novel approach to promoting the economic efficiency of fleet operators in road freight transport.

Zusammenfassung

Der europäische Strassengüterverkehr ist von zahlreichen Herausforderungen geprägt, wie dem zunehmenden Mangel an Lkw-Fahrpersonal, intensivem Wettbewerb und einem wachsenden Güteraufkommen im Lichte infrastruktureller Engpässe. Die Antriebswende hin zu emissionsfreien Fahrzeugen stellt langfristig einen weiteren Veränderungstreiber dar. Eine optimierte Lkw-Kapazitätsauslastung ist entscheidend für die unternehmerische Wettbewerbsfähigkeit und kann einen Beitrag zur Reduktion der ökologischen Auswirkungen leisten. Trotz des zunehmenden wirtschaftlichen Drucks auf Transportunternehmen sind Leerfahrten und nicht optimal ausgelastete Touren anhaltende Druckpunkte. Im Gegensatz zur zunehmenden Verbreitung der Sharing Economy in anderen Bereichen spielen Sharing-Ansätze im Strassengüterverkehr bislang keine nennenswerte Rolle.

Vor diesem Hintergrund untersucht die vorliegende Dissertation das Potenzial von Truck-Sharing als Ansatzpunkt zur Steigerung der ökonomischen Effizienz mit Blick auf bestehende Herausforderungen des Strassengüterverkehrs. Ziel der Arbeit ist es, ein zukunftsweisendes Konzept für Truck-Sharing zu entwickeln, welches gezielt die vorhandenen Ineffizienzen im Güterverkehr adressiert. Die Arbeit setzt mit Blick auf etablierte horizontale Kooperationen und die Sharing Economy in der unternehmerischen Praxis an und folgt einem gestaltungsorientierten Ansatz. Der Forschungsrahmen umfasst drei Untersuchungsbereiche: (1) die Exploration von Potenzialen und zukunftsgerichteten Anwendungsfällen, (2) die Untersuchung von zugrundeliegenden Treibern und Hürden und (3) die Ableitung von Implikationen und Handlungsempfehlungen. Der Forschungsprozess folgt einem explorativen Mixed-Methods-Design und verknüpft hierbei den Engaged Scholarship-Ansatz mit der Backcasting-Methodik. Als Forschungsmethoden werden Fallstudien, Experteninterviews und eine Online-Umfrage eingesetzt.

Das Ergebnis dieser Arbeit ist eine zukunftsorientierte Konzeptionalisierung von Truck-Sharing, die in der Umsetzung zur Reduzierung von Standzeiten von Güterkraftfahrzeugen und zur Erhöhung der betrieblichen Flexibilität beitragen kann. Ausgehend vom Status quo werden funktionale und wahrnehmungsbedingte Barrieren aufgezeigt, die es aus unterschiedlichen Blickwinkeln zu adressieren und durch Maßnahmen auf organisatorischer und individueller Ebene zu überwinden gilt. Verbunden mit Lösungsansätzen legt diese Dissertation den Grundstein für das Sharing von Güterfahrzeugen zwischen Transportunternehmen, als neuen Ansatz zur Förderung der wirtschaftlichen Effizienz von Flottenbetreibern im Strassengüterverkehr.

Contents

Abbreviations

AVE	Average variance extracted
bn	Billion
CEP	Courier express parcel service
e.g.	Exempli gratia (for example)
et al.	Et alia (and others)
FTL	Full truckload
ICT	Information and communication technology
IU	Intention to use
KMO	Kaiser-Meyer-Olkin
KPI	Key performance indicator
LSP	Logistics service provider
LTL	Less-than-truckload
M	Mean
mio	Million
OEM	Original equipment manufacturer
PEU	Perceived ease of use
PO	Psychological ownership
PTC	Perceived truck conditions
PU	Perceived usefulness
RQ	Research question
SD	Standard deviation
TAM	Technology Acceptance Model
TCO	Total cost of ownership
TS	Truck-sharing

List of Figures

List of Tables

1 Introduction to the Research on Truck-Sharing in Road Freight Transport

This research examines the roles and outcomes of truck-sharing (TS) in road freight transport. This first chapter outlines road freight transport challenges and connects the sectoral circumstances to TS as a solution approach. On the basis of a managerial perspective, a bridge to theoretical relevance is built. The research topic is investigated from a normative perspective, in which TS is considered a supplementary solution to overcome current and future challenges in efficient road freight transport. Research objectives and questions are derived in Sect. 1.3, according to managerial challenges and theoretical shortcomings. The chapter closes with an overview of the thesis structure.

1.1 Current Challenges in Efficient Road Freight Transport

Efficiency is central to measuring the performance of logistics networks and ensuring the economy's supply security. In road freight transport, the capacity utilisation of trucks indicates the efficiency of transport resource use. Although maximising capacity utilisation is relevant, and in the interest of both logistics service providers (LSPs) and the sectors reliant on road freight transport services, road freight transport is increasingly discussed in terms of sustainability. Because of transport-related emissions, freight transport has become central to public debates regarding efficiency. Reducing empty mileage and increasing truckload factors can lessen the environmental impact of freight transport and contribute to politically determined climate goals. According to analyses of fleet utilisation across Europe, approximately one in three truck trips is performed empty with any load, and the utilisation of trucks filled with loads ranges from

L. Häberle, *Truck-Sharing in Road Freight Transport*, Supply Chain Management,
https://doi.org/10.1007/978-3-658-48754-6_1

approximately 50% to 60% and has not significantly improved over time (EEA, 2010; KBA, 2023; BFS, 2023a).

Established instruments for transport optimisation—such as telematics systems, transport management systems, route optimisation software, cooperation between companies, and relying on freight marketplaces to balance out supply and demand on the spot market—support truck utilisation but cannot fully overcome operational challenges. Despite these widespread standard measures, empty mileage and inefficient use of transport capacity—and unintended drawbacks such as traffic congestion on roads—remain an ongoing challenge for LSPs and society. In practice, threshold levels of empty runs and utilisation factors below 100% are usually inevitable, because of the high complexity of transport optimisation associated with multiple restrictions, e.g., from the customer side, locations, unpairing of loads and routes, and vehicle specifications. In light of persisting operational challenges and evolving change drivers, a question arises as to which approaches are beneficial in addressing the problem.

Modal split and increasing road freight volumes

Road is the dominant mode of transport in European freight transport. In the EU-27 countries, the share of road transport in land transport is 74.6%, while rail has a share of 16.4% (European Commission, 2023). In Germany, the largest economy in the EU, road transport has a modal split share of 71.9%, in line with the EU average. Despite the greater importance of rail in Switzerland, particularly in transalpine freight transport, road transport is dominating the modal split with a 62.5% share in 2022 (BFS, 2023b).

The dominance of road transport in Europe has not diminished in recent years; whereas freight transport in the EU increased by 24.7% (+1.4% annually) between 2000 and 2021, freight transport by road increased disproportionately, by 38.6% (+1.6% annually), during this period (Eurostat, 2023).

Traffic forecasts suggest that the volume of freight traffic will increase further. According to Transport Outlook 2050, further freight traffic growth of 31.3% (0.9% annually) is expected for Switzerland by 2050 with respect to the base year 2017 (ARE, 2022). According to a long-term transport forecast (BMDV, 2024), the German Federal Ministry for Digital and Transport expects freight transport volumes to increase by 46% between 2019 and 2051. The share of road transport is expected to rise to 77.5% by 2051, because of a decline in bulk goods, such as coal and mineral oil, and a shift in shipment structure toward road transport, and away from rail and inland waterways.

Increasing freight volumes pose infrastructural challenges. The rising number of vehicles on roads is likely to further constrain infrastructure, which is already currently heavily utilised. This development is reflected in traffic congestion on roads.

Traffic congestion
Whereas the congestion hours on national roads in Switzerland reached a new all-time high in 2023 (with a 22.4% increase with respect to 2022, to 48,807 hours; ASTRA, 2024), the trend in congestion hours in Germany has also increased over time, although the record level from the pre-COVID-19 year 2019 was not reached again by the end of 2023 (ADAC, 2024). Approaches to mitigate the increase in congestion hours include expanding infrastructure at bottlenecks or decreasing overall traffic intensity. Infrastructure expansion is increasingly facing social resistance across Europe. Simultaneously, in light of constrained public budgets, infrastructural renovation backlogs—such as those in Germany, which have been highlighted by 21 leading business, transport, and logistics associations (Deutsches Verkehrsforum, 2024)—limit the scope for reducing congestion through expansion measures.

Thus, decreasing, or at least mitigating increases in, the number of vehicles on roads is crucial to ensuring a functioning transport infrastructure in the future.

Political and business challenges
A shift in the modal split in favour of road transport and overall growth in freight transport would increase freight transport by road above current levels. This shift poses infrastructural challenges at a political level, given that substantial financial expenditures for infrastructure maintenance and expansion will be required to transport goods by road in the future efficiently. However, additional challenges exist beyond the political level. Market participants in road freight transport face risks and specific restrictions to their competitiveness, some of which could intensify in the future.

Margin pressure in the logistics industry
The European logistics market is considered highly competitive, thus leading to low and sometimes even negative profit margins (Schwemmer, 2018). Although few data on profitability are available regarding logistics and specifically the road freight market, because of the dominance of small market participants, available data and market observations provide a consistent picture: the road freight market is considered a low-profit market across Europe (Schwemmer, 2018; Reimann, 2022). According to long-term observations by the German Federal Office for Goods Transport (BAG), the EBIT margins of market participants in road freight transport are predominantly between 1% and 3%. According to a historical survey

by the industry association International Road Transport Union (IRU, 2006), the average profitability of IRU members from EU-15 countries was only 0.6% in 2004.

Daher, Hausmann, and Wölfel (2022) have reported EBIT margins between 1% and 11% among the top 20 forwarders worldwide with public data, and 5% on average. These findings are in line with analyses by Schwemmer (2018), who has identified higher profit margins among large German LSPs with annual revenues exceeding 2 billion € (4.6%) than small companies with annual revenues below 20 million € (2.0%). These findings indicate that large players are more profitable than smaller ones. However, notably, margin differences usually result from logistics segments other than road freight transport (Table 1.1), and many small niche players are more profitable than segment averages indicate. A company-level view of the profitability of individual logistics companies does not provide pure information about profitability in the freight transport market, because companies often operate several business units. In the relevant full-truckload (FTL) and less-than-truckload (LTL) market, even large market participants such as Schenker or Kuehne + Nagel have low profit margins up to 2% (Paulsen, 2019). Covering the German market, Schwemmer (2018) has reported industry-wide average margins of 2.8% for FTL and 2.0% for LTL (Table 1.1). These market segments are below average and are among the segments with the lowest profit margins in logistics.

Table 1.1 Average profitability of logistics service providers by market segment in Germany in 2017[1]

Market segment	Market volume in bn €	EBIT margin
Bulk and carrier	26,774	7.3%
Terminal operations	4,200	13.8%
Special logistics services	12,689	3.1%
1. Consumer goods	6,095	2.3%
2. Tank and silo	3,751	3.2%
3. Other	2,843	4.0%
Contract logistics	34,460	2.7%
FTL	11,473	2.8%
CEP	32,113	1.5%
LTL	16,648	2.0%
Air and sea	31,211	2.2%
Overall	**169,838**	**3.3%**

[1] Schwemmer (2018).

In this regard, attention has been drawn to the development of customer-side transport demand, which generally is scarcely influenced by LSPs. The demand for freight transport is derivative, in that it is derived from economic and social needs to exchange goods (Stölzle, Häberle & Schöder, 2024). LSPs can plan demand to only a limited extent and therefore must react to fluctuations over time. Companies' transport capacities can be adjusted to only a limited extent in the short term; consequently, balancing transport supply and demand to maximise truck utilisation is a permanent operational challenge.

Thus, from a financial perspective, unlocking efficiency measures to improve the profitability of low-margin road freight transport is in the interest of LSPs operating their own transport capacity.

Increasing truck driver shortages across Europe

In recent years, the recurring imbalance between transport demand and supply has manifested as a persistent bottleneck in transport capacity (Kille, Schmidt, Stölzle, Häberle & Rank, 2023). One long-term driving force underlying this development is socio-demographic changes, which represent a further operational challenge for LSPs. The truck driver shortage is a substantial problem in freight transport (Srinivas, Ramachandran & Rajendran, 2022).

In its *Europe Truck Driver Shortage Report*, the World Road Transport Organisation IRU has stated that more than 230,000 truck driver positions were unfilled in 2023 in Europe, and has predicted an increase to 745,000 by 2028 if no substantial measures are taken (2024). Demographic changes are the main driver of this change. What is reported for Europe has been analysed in detail for the German market by Kille, Schmidt, Stölzle, Häberle, and Rank (2023). According to that study, the truck driver shortage in Germany alone was 56,000 drivers in 2022, considering that approximately 560,000 people were employed in the profession (BALM, 2024), 10% of all job positions were unfilled in this profession—a percentage higher than that in other typical shortage occupations. Because of demographic changes and a lack of young truck drivers, a net loss of approximately 20,000 truck drivers per year was assumed, given that 35,000 job exits due to retirement were offset by only 15,000 new entrants.

Consequently, competition among companies for truck drivers is intensifying. The shortage improves the negotiating position of truck drivers, who can choose employers providing higher salaries and more attractive working hours. Because truck drivers spend a large part of their working time in the vehicle, LSPs are increasingly willing to configure trucks according to the wishes of their truck drivers, such as by selecting preferred brands and adding special equipment.

Companies without sufficient truck drivers face difficulties in staffing vehicles. Truck downtimes therefore increase, because truck transport cannot be performed without qualified truck drivers.

Thus, the prevailing and even intensifying truck driver shortage is increasing the pressure on efficient truck deployment, in light of rising freight transport volumes and the continued modal split dominance of road transport. In addition, given truck drivers' strong negotiating position, transport companies must offer attractive working conditions and prioritise the needs of truck drivers to a greater extent than they did in the past.

Adoption of autonomous truck operations not yet foreseeable in Europe

Autonomous driving is often discussed as a solution to the challenge of truck driver shortages in the industry that additionally may provide possible benefits such as reducing operating costs and increasing safety (e.g., Wrenn, 2017; World Economic Forum, 2021; Schuster et al., 2023). In principle, autonomous driving has the potential to mitigate the truck driver shortage if truck drivers can be fully or partially replaced. However, doing so would require a high level of automation. According to the definition of levels of automation from the Society of Automotive Engineers, such automation corresponds to at least level four (SAE International, 2021), defined as self-driving vehicles, in the sense that a truck is operated without any human control or supervision (level five) or under specific conditions (level four) (Herrmann, Brenner & Stadler, 2018). These two levels potentially allow the driving task of a truck driver to be fully replaced.

Despite technological progress and the promising benefits of autonomous truck operations, the actual adoption timelines in commercial road freight transport depend on multiple factors, and remain vague and thus scarcely predictable (Simpson, Mishra, Talebian & Golias, 2019; Schuster et al., 2023), possibly because of numerous challenges. Economic considerations; social acceptance; and the state of regulatory, technologic and infrastructure readiness are prerequisites for use of autonomous trucks (BCG, 2022; Graf & Anner, 2021). For example, autonomous trucks require adequate infrastructure and infrastructure quality to enable consistent and reliable operation. Environmental conditions differ broadly and are likely to hinder widespread autonomous truck adoption. Mixed-traffic environments might provoke software failures or unexpected situations, thereby negatively affecting trust and regulatory support. Hence, low-traffic long-haul routes are considered most likely to operate autonomously (Slowik & Sharpe, 2018), whereas handling in the first and last miles of a route is considered more complex and is expected to require human truck drivers in the future. Dense road systems, high traffic congestion, and shorter road transport distances suggest that adoption may occur later in Europe than in North America.

Thus, although autonomous driving is a promising technological innovation, this work is based on an assumption that autonomous driving will not find substantial application in European road freight transport in the foreseeable future until 2035, and will trigger significant industry changes. However, adoption may differ for another change driver, alternative drives.

Accelerating powertrain transition to achieve sustainable transport solutions despite existing hurdles
As a macro trend, sustainability targets, at a political level, drive change in the industry. The pressure to reduce emissions from freight transport is leading to a shift away from diesel-powered drive systems and towards alternative drive systems. Electric vehicles are considered the most promising possibility for decreasing carbon emissions in freight transport in the long-term (Alarcón, Mac Cawley & Sauma, 2023). Electric vehicle adoption in road freight transport is thus likely to dynamically increase over time.

However, engine transition started later and takes longer for heavy vehicles than cars, and new vehicle registrations for heavy vehicles are lagging behind those for light vehicles in freight transport (IEA, 2024). Alternative drive systems currently do not significantly contribute to vehicle fleets. As of 1 January 2024, the number of diesel tractor units among utility vehicles was 2.2 million in Germany. In contrast, only 1244 battery electric vehicles and 4,095 natural gas vehicles were registered (KBA, 2024). The numbers for medium-duty lorries are slightly higher, including at the European level (Werwitzke, 2024). Examination of new vehicle registrations has indicated that 5% of all newly registered medium-sized trucks in Europe in 2023 were electric vehicles, and less than 1% were heavy trucks. Some national markets are developing more dynamically. For example, the proportion of newly registered trucks with a payload exceeding 3.5 tonnes was 10% in Switzerland in 2023 (Friedli, Zacharias & Häberle, 2024).

The delayed transition in heavy goods transport is a result of specific challenges in freight transport, particularly long-haul transport; central barriers to the adoption of electric trucks include short ranges, lack of charging infrastructure, and high investment costs (Alarcón, Mac Cawley & Sauma, 2023; Qasim & Csiszár, 2021). For example, purchase prices for electric trucks are currently two or three times those of equivalent trucks with combustion engines. Therefore, financing is challenging, regardless of whether a truck is purchased or leased, because truck deprecation costs are the third largest cost block faced by transport companies, after truck driver wages and fuel costs (BGL,2024).[2] The transition to alternative drive systems is therefore cost-intensive for fleet operators. The cost block of truck acquisition costs will increase as alternative drive systems are adopted. In a TCO analysis, the advantages of electric trucks stem from lower operating costs, which offset the higher initial purchase costs over the long term. Higher truck costs are associated with rising opportunity costs for parked vehicles in a firm's own fleet, which cannot be utilised in situations of low transport demand. From an efficiency perspective (Sect. 2.1), vehicle downtimes are undesirable and become increasingly important with higher vehicle purchase costs.

Thus, as the transformation to alternative drive systems, particularly electric trucks, progresses, truck downtimes will become an increasingly important performance indicator for truck fleets. Reducing downtimes can serve as an efficiency lever with respect to the acquisition costs of electric trucks.

Unlocking efficiency potential remains a continuing challenge
In light of changes in the industry, efficiency is notably not a new phenomenon for transport companies but instead is a core component of operational activities. Optimising capacity utilisation in transport networks of LSPs and consolidation LSPs can contribute to both the economic and environmental performance of transport operations, and thus of their customers, as described by Lam and Dai (2015).

Cost-effective service provision fosters competitive advantages in the competitive transport market. Several strategies have been established by transport companies to achieve this goal. Internal optimisation in the sense of load consolidation is a standard. Most internal planning processes of LSPs are technologically supported, for instance by route planning software. Engaging in cooperation (Sect. 2.2) with other transport companies can balance out the fluctuations and load imbalances in a company's own network. Transport companies use freight exchanges, which serve as marketplaces for exchanging loads. Joint route planning in fixed networks and bilateral cooperation on specific routes are also practiced (Verdonck, Caris, Ramaekers & Janssens, 2013). Nonetheless, empty running and underutilised trucks remain challenges (KBA, 2023; BFS, 2023a), thus prompting a search to improve efficiency in road freight and foster sustainable freight transport. Which further approaches are suitable for addressing these issues? Insights from the sharing economy may offer opportunities.

[2] Depending on the truck deployment profile and national tax policies, companies' the cost structures of companies may differ in individual cases.

Truck-sharing addresses economic, ecological, and social challenges

The sharing economy has received broad attention in recent years, in terms of leveraging the sharing of underutilised assets to improve efficiency and sustainability (Hossain, 2020). Adoption of the sharing economy has occurred in various contexts, such as accommodation (Airbnb) and mobility (Uber). Observed phenomena of collaborative consumption of goods and services involve activities such as swapping, trading, or renting, thereby offering the asset owner a higher utilisation rate (Hamari, Sjöklint & Ukkonen, 2016). Car sharing has emerged as a commonly used and widespread form of sharing in human mobility. Although the idea of sharing is not new in freight traffic, the prior focus has been on horizontal cooperation sharing vehicle capacity (Cruijssen, 2006; McKinnon, 2000) rather than on exchanging physical assets, namely trucks.

TS refers to a flexible, demand-oriented exchange of resources between fleet operators. The physical exchange of assets is at the core of the sharing concept, thereby making underutilised trucks accessible to other market participants in road freight transport. This pooling of capacity across companies is aimed at balancing demand peaks with a reduced total number of trucks, thus enhancing the efficiency of resource utilisation. Depending on the application scenario, this exchange can occur with or without the inclusion of driving personnel.

The previously outlined cluster of challenges in road freight transport establishes the boundary conditions and foundational context for investigating the phenomenon of TS. This research was aimed at addressing multiple objectives across economic, ecological, and social dimensions, as follows:

- *economic:* efficient transport (costs)
- *ecological*: sustainable transport (emissions)
- *social*: traffic congestion

The sharing economy, with its focus on optimising the deployment of underutilised assets, supports the idea of TS increasing the overall loading of a truck by reducing empty mileage or only partially loaded trucks (Islam, Olsen & Daud Ahmed,2013). Cooperation among companies at the same level of the supply chain, similarly to capacity sharing, is aimed at decreasing costs across the industry and increasing responsiveness to environmental changes (Matusiewicz & Ksiazkiewicz, 2023). Islam and Olsen (2014) have examined TS collaborations among container road carriers presenting efficiency gains through shared logistics. Despite these theoretical insights, the extent of practical application of TS remains very limited (Vendela, 2017), and the exchange of physical resources such as trucks between parties is under-researched in this context.

1.2 Research Objective and Questions

This research is set against a backdrop of efficiency challenges in road freight transport, comprising growing freight volumes, infrastructural constraints, and needs for sustainable transport solutions in Europe. This work further addresses the persistent issue of low profit margins in light of cost increases in the logistics industry and intensifying pressure to manage truck driver shortages.

By leveraging insights from the sharing economy and horizontal cooperation in the industry, this thesis addresses TS as a solution approach. Specifically, the research was aimed at investigating the role and application potential of TS to increase efficiency in road freight transport. Furthermore, this thesis provides a foundation for establishing a new operational tool for use in transport companies' freight operation planning.

Because TS, as a new field of research, lacks consistent terminological usage in the academic discourse, this thesis is aimed at fundamentally defining the term 'truck-sharing' and to establish a shared conceptual understanding among researchers and practitioners.

Through analysing challenges related to the utilisation of transport capacity and the limitations of existing cooperative approaches, including freight marketplaces and bilateral agreements, this thesis explores use cases for transport companies to identify the specific application potential of TS in practice.

Beyond contributing to the scientific debate, this thesis may be impactful for practitioners. Herein, drivers and barriers to TS adoption in practice are examined by involving key stakeholder groups. In this way, tangible recommendations are derived for implementation in practice.

Thus, this research adopts a collaborative engaged scholarship approach to integrate multiple stakeholder perspectives. Because of the novelty of the topic, an exploratory mixed method design s applied by combining explorative, qualitative interviews among practitioners in road freight transport with a subsequent survey conducted among transport companies and truck drivers. This process integrates different perspectives and insights from key stakeholders entrenched in real-world problem context.

This research contributes to the academic discourse on sustainable logistics and offers practical insights for industry stakeholders. It is aimed at bridging the gap between theoretical concepts and practical application of TS, thereby supporting the development of more efficient, sustainable, and resilient road freight transport networks.

According to the research objectives outlined above, three research questions (RQs) are investigated in this thesis:

RQ1: What **potential** and future **use cases** contributing to increased efficiency are emerging for TS in the context of road freight transport?

RQ2: What are the key **drivers** of, and **barriers** to, TS in road freight transport?

RQ3: What **implications** and **recommendations** can be drawn to leverage the adoption of TS in practice?

1.3 Structure of the Thesis

This thesis is divided into six chapters, each directed at answering the RQs. Fig. 1.1 illustrates the structure of the thesis.

After the introduction in this first chapter, *Chap. 2* presents theoretical and conceptual fundamentals, grounded in efficiency measures regarding truck utilisation. The intersection of horizontal cooperation in road freight transport and the sharing economy is outlined, to delineate conceptual fundamentals of TS to establish a shared conceptual and terminological understanding.

Chapter 3 describes the research approach. Sections. 3.1 and 3.2 describe the research paradigm and theoretical approach to investigating TS. Subsequently, Sect. 3.2 presents the research design.

Chapter 4 presents the empirical research and findings. Initially, the potential applications of TS are investigated through a case study-based status quo analysis of fleet utilisation. Section 4.2 identifies fundamental drivers and barriers, and culminates in a series of application-oriented implications and recommendations.

Chapter 5 summarises the thesis's research findings and addresses its contributions and managerial implications. Finally, limitations and opportunities for further research in this field are discussed.

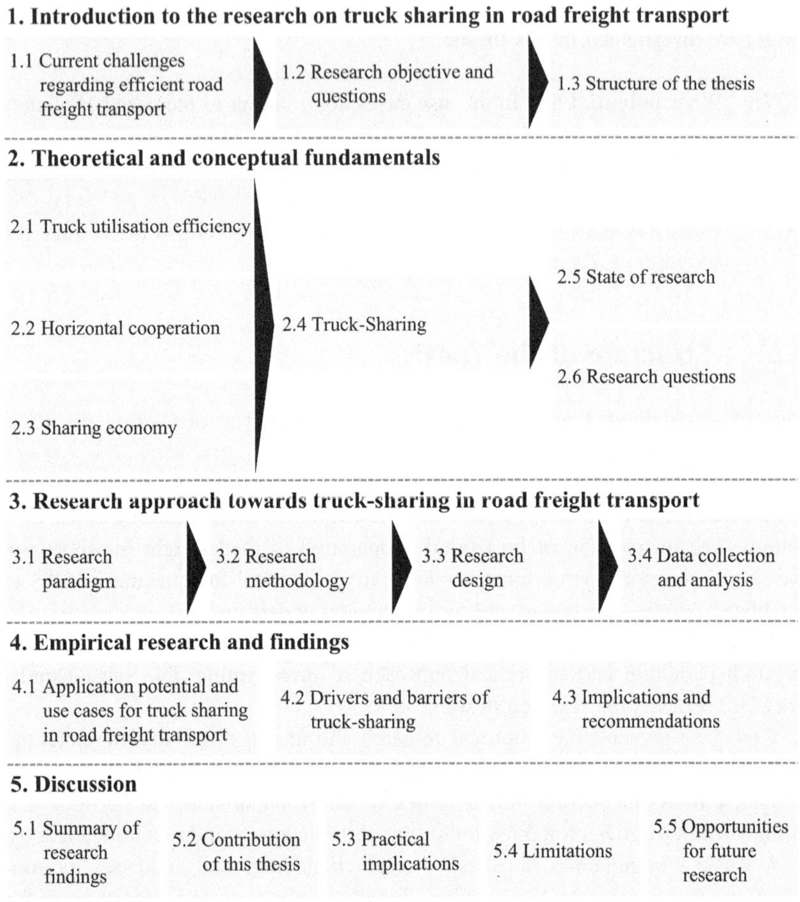

Fig. 1.1 Outline of the thesis structure

2 Theoretical and Conceptual Fundamentals

Organising freight transport efficiently is a key objective of LSPs. Various approaches may be used to define and measure efficiency, thus necessitating an explanation of practical challenges in utilising trucks efficiently and where exactly TS might be applied. This chapter therefore first provides an overview of understanding and measuring the efficiency of transport operations, then introduces relevant literature streams. This thesis focuses on the business aspects of the efficiency of TS. The objective is to optimise the use of resources within companies by optimising existing transport capacities, reducing costs, and increasing profitability. The thesis therefore concentrates on the immediate benefits for companies involved, such as the reduction of operating costs and maximisation of truck utilisation. Economic and environmental efficiency, such as the overall economic distribution of resources and decreasing the environmental footprint, are considered positive side effects of the successful implementation of an economically efficient TS model. However, these aspects are not the focus of this study and follow the economic optimisation.

Cooperation across company boundaries is not a new phenomenon but is an essential part of logistics networks. What types of cooperation exist, and for what purposes are they used? Moreover, how does TS fit into existing cooperative approaches? Because fleet operators typically cooperate with other fleet operators, at the same level of the value chain, the literature on horizontal cooperation is a central literature stream.

Second, the sharing economy is a literature stream foundational to the concept of TS. The sharing economy, with its focus on maximising the utilisation of underutilised assets, supports the idea of TS increasing the overall loading of a truck by completely reducing fully empty or partially loaded trucks (Islam, Olsen & Daud Ahmed, 2013). Despite theoretical insights from the sharing economy, practical application of TS remains limited, according to Vendela (2017),

L. Häberle, *Truck-Sharing in Road Freight Transport*, Supply Chain Management,
https://doi.org/10.1007/978-3-658-48754-6_2

and the exchange of physical resources such as trucks between parties is under-researched in this context, partly because of the inconsistent use of the term 'sharing' specifically for trucks. Sharing a truck usually focuses on sharing the truck's capacity rather than the vehicle itself. Vehicle sharing in mobility, only covering the sense of a physical resource exchange, is found and observable in relation to cars but not yet trucks in road freight transport. Therefore, the available literature is restricted because it does not relate specifically to trucks.

Given these observations, a literature review on relevant research streams of efficiency related to freight transport, horizontal cooperation in logistics, and the sharing economy is provided, with a focus on findings relevant to establishing TS. This review provides a comprehensive understanding of the efficiency challenges addressed by sharing trucks as well as the transfer potential of established forms of cooperation in logistics and sharing concepts in other fields. These streams are merged to identify relevance to TS, as illustrated in Fig. 2.1. In this context, TS is not a direct consequence of, and is not present solely after, the fulfilment of elements from all three research streams. Instead, the model serves as a guide for the chosen theoretical streams, each of which contributes to the purpose of this thesis. Depending on the research focus, research on TS may also draw on other literature streams.

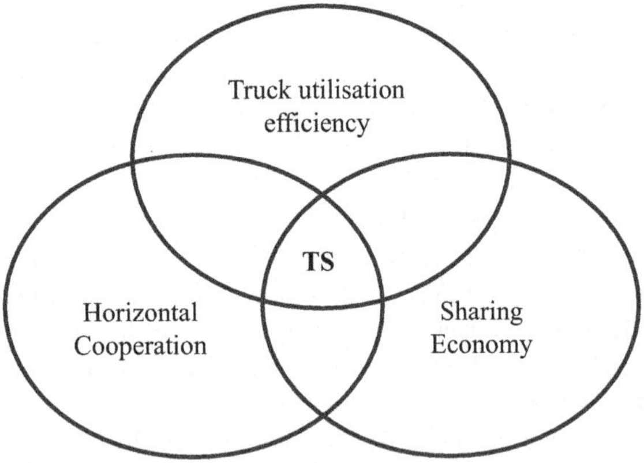

Fig. 2.1 Truck-sharing in light of transport efficiency, horizontal cooperation, and the sharing economy

2.1 Truck Utilisation Efficiency

The starting point for the consideration of efficiency is the role of freight transport as a core component of functioning logistics and, more generally, functioning supply chains. The term 'supply chain management' was introduced in 1982 by Oliver and Webber (1982) to address complexity and the increasing competition between companies. This thesis follows the understanding of Christopher (2011), who defines supply chain management as *"the management of upstream and downstream relationships with suppliers and customers in order to deliver superior customer value at less cost to the supply chain as a whole" (p. 3).*

Efficiency, derived from this definition, is considered a key parameter central to the design of supply chains. These supply chains can be structured according to service-specific subsystems analogously to the logistics chain. Transport is one of the three core activities, which also include handling and storage (Pfohl, 2003). The transport of goods is crucial for the seamless operation, efficiency, cost structure, and adaptability of the entire supply chain. Transport is the element connecting the various stages of the value chain, such as procurement, production, warehousing, and distribution. Transport enables the flow of raw materials, semi-finished, and finished goods from suppliers to manufacturers and ultimately to the end customer.

Efficient freight transport is in the interest of both corporate entities, such as shippers and transport companies, and policymakers, because of its potential to reduce logistics costs and minimise traffic-related environmental impacts, such as greenhouse gas emissions. Increasing transport efficiency contributes to lower traffic congestion, decreased travel distances, and reduced fuel consumption. On the one hand, optimising transport efficiency is an integral part of daily business operations for transport companies, to enable them to compete in the competitive market environment. On the other hand, the establishment of more competitive and resource-efficient transport systems has been described as an approach to achieving the European Union's climate change objectives (European Commission, 2011). Thus, utilising available freight transport resources is considered relevant from both an environmental perspective and an operational perspective (Pahlén & Börjesson, 2012). In line with this development, Zhao, Ke, Zuo, Xiong, and Wu (2020) have described increasing recognition of sustainability issues in transport research in recent years. Because of restricted data availability and complexity, little knowledge of actual resource utilisation has been obtained (Pahlén & Börjesson, 2012).

According to Piecyk (2010), three dimensions can be used to measure the efficiency of transport. First, the *load factor* is defined as the proportion of actually utilised capacity of a vehicle measured against its total available capacity. Second, *resource efficiency* refers to the utilisation of individual vehicles in a transport network concerning empty mileage and vehicle downtimes. This measure is related to information systems and planning tools for vehicle routing. Third, *system efficiency* considers the network structure of the transport system, including the location of logistical facilities aligned to meet customer demands.

Those three dimensions vary according to the decision level and planning horizons. Whereas the load factor is spatially related to the loading space of a transport unit itself, the second dimension, resource efficiency, adds a temporal and distance component, and the third dimension, system efficiency, covers an entire fleet or transport network. Increasing the load factor of vehicles and optimising vehicle efficiency are decisions made on an operational level, whereas system efficiency is considered to be at the strategic decision level and implies a long-distant planning approach. Focusing on the factors that can be influenced at the operational level in daily freight management, McKinnon (2010) has defined five utilisation measures comprising the load factor and vehicle efficiency:

1. **Weight-based load factor**: ratio between the actual weight of the transported goods and the maximum weight that could have been loaded on a vehicle
2. **Tonne-km load factor**: ratio between the actual tonne-kilometres and the maximum tonne-kilometres that could have been transported if a vehicle had been fully loaded on a tour
3. **Volumetric load factor**: ratio between the actual cubic space occupied by the load in the vehicle and the maximum cubic space that could have been loaded
4. **Deck area coverage**: proportion of the area of the vehicle floor covered by a load, in a two-dimensional representation
5. **Empty running**: proportion of a truck's travelling distance run empty, according to either the number of trips or the distance travelled in kilometres

These five measures are distinct and can be analysed independently of one another. However, from a practical viewpoint, the significance of each measure can be assessed only in a wider context. For example, a truck transporting heavy cargo might reach the maximum weight-based load factor but not be fully utilised in terms of volume and deck area coverage. Therefore, from a business perspective, applying assessment approaches that consider multiple parameters is useful.

Examining only the utilisation of loaded trips would ignore the empty kilometres travelled. Therefore, in my view, the interplay of several measures should be used to meaningfully utilise efficiency analysis.

Table 2.1 Results of selected studies and statistics on truck utilisation in Europe

Country	Measure	Utilisation	Source
UK	Volumetric (3)	52%	McKinnon & Ge (2004)
Sweden	Volumetric (3) Weight (1)	61% 64%	Pahlén & Börjesson (2012)
Germany	Weight (1) Tonne-km (2) Empty trips Empty running	56% 54% 38% 23%	KBA (2023)
Switzerland	Empty running	26%	BFS (2023a)

Empirical analyses of the capacity utilisation of heavy goods vehicles have indicated that the average capacity utilisation in commercial freight transport is considerably below the theoretical maximum of 100%. Table 2.1 summarises the results of selected studies and official statistics on truck capacity utilisation in Germany, Switzerland, Sweden, and the UK. Across those countries, the values for volume and weight-related capacity utilisation have indicated load factors between 52% and 64%. These values are in line with findings reported by Greening, Palmer, Piecyk and McKinnon (2015), who have identified load factor improvement potentials ranging from 30% to 50%. Regarding empty running share, public data for Germany and Switzerland have consistently indicated high proportions of empty kilometres, of 23% and 26%, respectively. The significantly higher proportion of empty trips (38%) than empty kilometres is explained by short distances between an unloading point and the next loading point being recorded as one empty trip. Short empty trips are offset by longer load trips. The proportion of empty kilometres travelled is therefore a better measure than looking at the proportion of empty kilometres travelled.

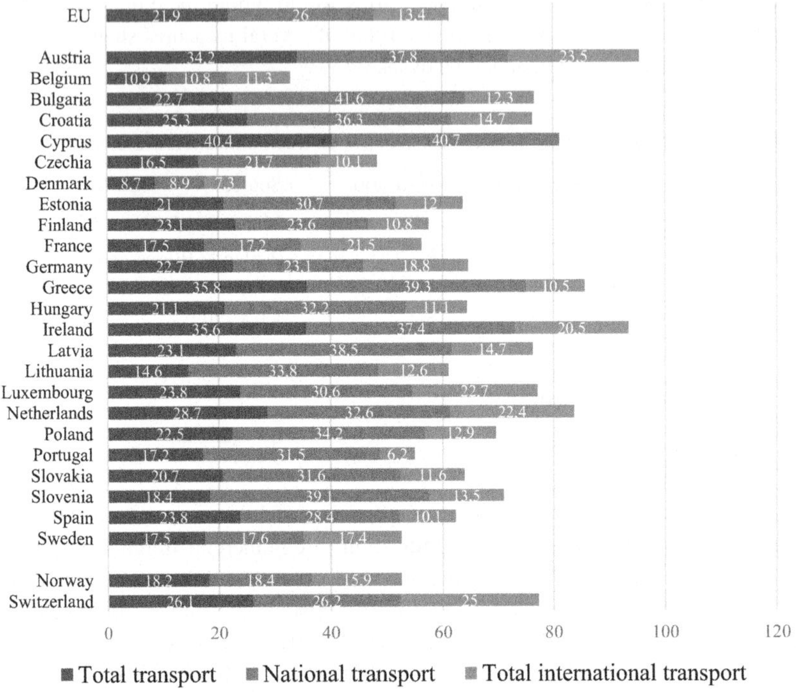

Fig. 2.2 Empty road freight traffic in European countries by type of operation as percentage shares in vehicle-km in 2022[1]

Utilisation differences are observed depending on the type of truck operation. Domestic road freight is associated with a much higher rate of empty running than international road freight (Fig. 2.2). At the European level, 13% of vehicle kilometres in international road freight transport are run by empty vehicles, whereas at the national level, empty mileage values are higher on average, at 26%, but large deviations are observed among individual states. Rates of empty running have remained relatively constant across countries since 2010.

Utilisation rates of trucks in road freight transport appear low with respect to the theoretical maximum utilisation of 100%, because of a variety of operational, logistical, and market-driven factors and restrictions. Mapping these limitations is

[1] Eurostat (2023). Italy, Malta and Romania are excluded from the analysis, because data are not reported.

crucial to develop strategies to enhance the efficiency and thus the sustainability of road freight transport. Table 2.2 provides an overview of limiting factors faced by transport companies regarding the optimisation of freight dispatching.

A fundamental constraint is the extensive range of service profiles in road freight transport, coupled with the restrictions that apply to specific types of goods. In the absence of these constraints, optimisation would be a simpler, or a feasible, undertaking. For instance, shipments are not always compatible with one another for various reasons. Consequently, goods that fit into a tour in terms of space and time might not be able to be transported in the same truck because of other restrictions. For example, foodstuffs frequently cannot be transported with hazardous goods. Requirements for narrow temperature zones can also prevent some pharmaceuticals from being transported with other pharmaceuticals.

Table 2.2 Factors limiting the utilisation rate of trucks in road freight transport[2]

Category	Factor	Description
Unpaired freight flows	Directional imbalance	Trucks may return empty or only partially loaded, owing to uneven freight flows in different directions.
	Regional differences	Economic activities and consumer behaviours are unevenly distributed geographically, thereby leading to imbalanced freight patterns.
Demand structure	Small shipment size	Many shipments do not fill a truck's capacity, thus leading to less-than-truckload and general cargo operations.
	Variable demand	Seasonal trends, economic cycles, and daily or weekly fluctuations lead to inconsistent demand.
Operational constraints	Scheduling issues	Synchronising pickup and delivery times can be challenging, e.g., because restricting service level agreements may cause delays and underutilisation.
	Regulatory restrictions	Regulations regarding truck drivers' driving hours and mandatory rest periods limit optimisation of routes and schedules.

(continued)

[2] Based on Stölzle and Häberle (2024).

Table 2.2 (continued)

Category	Factor	Description
	Infrastructure bottlenecks	Traffic congestion limits planning capability; consequently, delays or buffer times may limit effective truck utilisation.
Route planning	Suboptimal routing	A lack of sophisticated route optimisation technologies leads to inefficient use of capacity and additional kilometres
	Geographical barriers	Physical barriers such as mountains and rivers create longer, less direct routes.
Economic pressure	Cost minimisation focus	Competitive pressure to minimise transport costs can override efforts to maximise load capacity.
	Market fragmentation	The fragmented transport market dominated by small carriers lacks the scale for efficient load maximisation.
Customer requirements	Just-in-time	Timely delivery demand can lead to trucks not being fully loaded, to meet specific delivery windows.
	Service level agreements	High service level expectations are reflected in contractual agreements; e.g., delivery times conflict with high utilisation rates, because they restrict the carrier's planning capability.
	Variable demand	Shippers tend to favour flexibility over fixed quotas. Demand fluctuations from the shipper side force dynamic adaptation.
Freight specifics	Heterogeneous transport requirements	Combining freight types, such as dangerous goods and temperature-controlled freight, is limited, because of safety, security, or quality reasons.
	Means of transport	The choice of means of transport and handling requirements depends on the nature of the goods. Solid, liquid, and gaseous goods sometimes require specific equipment.

Examination of the range of factors reveals that utilisation limiting factors can have different effects. First, owing to freight transport's derivative character, volatility in the form of demand-side fluctuations can lead to trucks coming to a standstill and not being used for transport temporarily. Second, directional imbalances in freight flows can lead to empty runs when no suitable return load is available; therefore, a truck drives completely empty on the road. Third, in addition to these two prior effects, truck efficiency is decreased when trucks are loaded but not completely filled, e.g., when single shipments on a tour cannot be sensibly consolidated into a full truckload. On the basis of this consideration and McKinnon's (2010) measures of efficiency, the three dimensions are adopted as measures of transport efficiency in this thesis, as shown in Fig. 2.3. The primary measures *empty mileage* and *utilisation rate* refer to the running state of vehicles, whereas *vehicle downtime* is a secondary measure referring to standing vehicles.

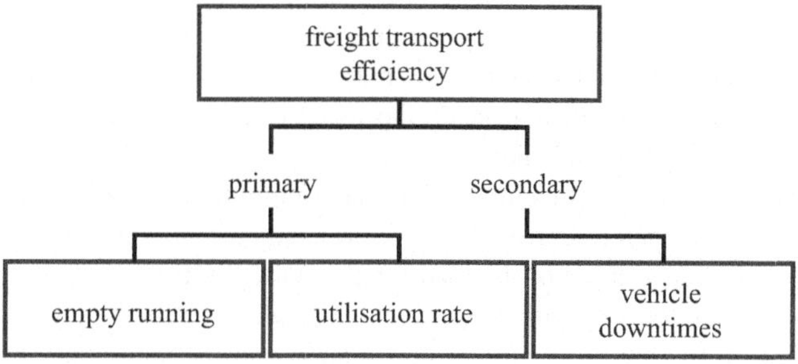

Fig. 2.3 Classification of freight transport efficiency measures

Transport companies rely on a range of standard approaches to address shipment restrictions and plan routes with the best possible capacity utilisation in the dynamic, complex environment of transport planning optimisation. In dispatching, transport management systems are at the heart of the planning, execution, and optimisation of transport processes. Transport management systems are complemented by route planning software—specialised software tools that facilitate the calculation of the most efficient routes by considering traffic, road conditions, and delivery priorities. Shipments are consolidated to optimise load space by combining general cargo and partial loads into full truckloads. Many companies rely on cooperation (Sect. 2.2) with other transport companies (e.g., regarding general cargo) and use freight exchanges to find suitable return freight or offer freight on the market if it does not match the routes. Freight exchanges are a widely used means of operating on the spot market to balance out fluctuations and address the backhaul effect (Arvidsson, Woxenius & Lammgård, 2013).

Despite the described potential and anticipated prediction of an improvement in capacity utilisation among transport (e.g., Dong, Akram, Andersson, Arnäs & Stefansson, 2021)—driven by the use of advanced technological solutions such as telematic systems, automated disposition processes, and real-time visibility to improve route and cargo space planning—no substantial improvements in truck utilisation rates can be identified in practice in recent years (e.g. KBA, 2023).

Political and entrepreneurial efforts to enhance efficiency in sustainable freight transport thus remain in place. If existing approaches do not sufficiently contribute to sustainability improvements, questions arise regarding further previously unutilised solutions, such as the sharing approaches examined in this research. This thesis considers the limiting factors presented in Table 2.2 as restrictions that LSPs must address in practice. TS is not aimed at eliminating these restrictions directly; instead, the focus is on identifying approaches to mitigate the resulting inefficiency.

Whereas the potential of technology in road freight transport is future-oriented, cooperations have been established for decades and currently are anchored in and characterise the functioning of the road freight market in Europe. The subsequent section addresses the topic of cooperation in relation to road freight transport, and focuses on the various forms of cooperation and the value that they add in road freight transport.

2.2 Horizontal Cooperation

More than 559,000 companies comprising many market participants are active in the European road freight transport industry (European Commission, 2023). The incremental liberalisation of the freight transport market in Europe, particularly in the 1990s, has led to increased competition between transport companies and across national borders, as a feature of the market. In view of the high intensity of competition and cost pressure, horizontal cooperation between transport companies is currently widely practised. Particularly in the general cargo market, many cooperations have been established to offer cost-effective, nationwide network coverage. In Germany, with the exception of *IDS Logistik* (1982)[3] and *NG.Network* (formerly *System Alliance*, 1985), all established general cargo

[3] The founding years of individual cooperations are indicated in parentheses.

cooperations were founded in the 1990s: *CTL* (1993), *Cargoline* (1993), *Palletways* (1994), *online Systemlogistik* (1995), *24plus* (1996), *Sim Cargo* (formed from *STAR* (1997), *ILN* (1998)), and *VTL* (1998). With *systempo* (1999) in Austria and *Cargo24* (2009) in Switzerland, one general cargo cooperation has been established in each of these two countries. The founding of *E.L.V.I.S.* (2006) for transporting partial loads and *PANECO* (2021) as the first pan-European network and 'cooperation of cooperations' indicated the expanding focus of cooperations.

The aim of Sect. 2.2 is to provide an overview of the forms of cooperation, followed by their objectives and framework conditions.

2.2.1 Types of Cooperation in Freight Transport

Collaborations are characterised by voluntary cooperation with participating companies maintaining their legal and economic independence (Berger, 2009). Engaging in coordinated actions enables achievement of shared objectives that would not be achievable if the actions were performed by companies individually.

According to Pfohl (2018), forms of cooperation in the transport industry are generally classified as inter-company and supra-company. Whereas inter-company cooperation involves direct collaboration between individual companies in the transport industry, cross-company cooperation involves coordination of tasks in a jointly organised institution or with outsourcing (Wohlgemuth, 2002). The latter form of cooperation is aimed at meeting the specific needs of the members (commercial cooperation) or bundling the interests of the partners (commercial-political cooperation) and thus goes beyond core company activities.

Inter-company cooperation in the transport industry takes a variety of forms. The criterion of the direction of cooperation is used to further categorise inter-company cooperation into *horizontal*, *vertical*, and *lateral*[4] *cooperation* (Fig. 2.4).

[4] The term diagonal is frequently used as a synonym for lateral cooperation (e.g., Pfohl, 2018).

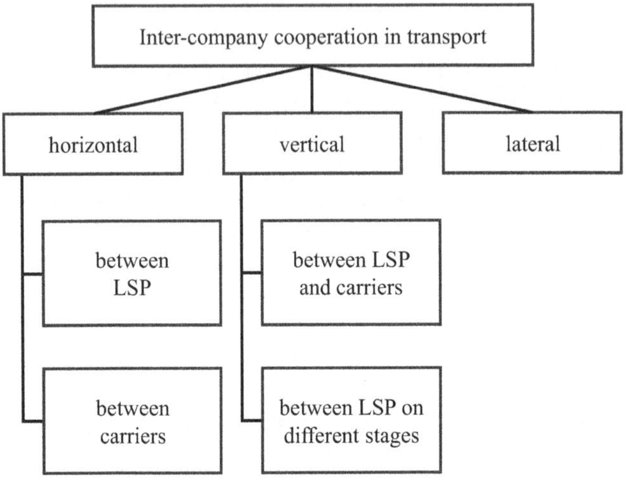

Fig. 2.4 Forms of inter-company cooperation in transport by direction[5]

Cruijssen, Dullaert, and Fleuren (2007) first defined horizontal cooperation as "*cooperation between two or more firms that are active at the same level of the supply chain and perform a comparable logistics function on the landside*" (p. 26). This definition implies cooperations among LSPs or carriers without involvement of other stages of the supply chain. Freight transport cooperations are usually a horizontal form of cooperation (Schneider, 2024). Schmoltzi and Wallenburg (2011) have found that most LSPs engage in one or multiple horizontal cooperations with other LSPs. In practice, a broad range of horizontal cooperations are found, including purchasing partnerships (between LSPs or shippers), logistics networks (e.g., in cooperations for general cargo and partial loads, as described at the beginning of this chapter), city logistics concepts, and marketplaces for freight and load space balancing. The advantages typically include joint improvements in performance or cost reductions. Depending on the configuration of resources, two fundamental types of relationships can be identified (Freichel, 1992). When the

[5] Based on Pfohl (2018).

resource profiles of the involved companies are aligned, the primary objective is leveraging existing economies of scale in the short to medium term. In contrast, horizontal relationships based on complementary combinations of resources are formed to develop new potential in the medium to long term. These relationships are particularly conducive to the development and utilisation of competencies.

Second, vertical cooperation involves stakeholders from different stages of the supply chain across logistics chains. The cooperation may involve customers or suppliers in upstream or downstream stages. Cooperations with other LSPs at other stages of the supply chain also fall under this category. Compared with horizontal cooperations, vertical cooperations between LSPs and shippers are more widespread (Renko, 2011; Basso, D'Amours, Rönnqvist & Weintraub, 2019), as reflected in an extensive body of literature investigating partnership structures and attributes (e.g., Stefansson, 2006, Gibson, Rutner & Keller, 2002), and performance drivers (Lambert, Emmelhainz & Gardner, 1999; Deepen, Goldsby, Knemeyer & Wallenburg, 2008).

Third, lateral cooperation combines horizontal and vertical synergies, and is usually intended to increase flexibility. Flexibility is also central to the stage model by Freitag, Kück, and Becker (2016), which classifies cooperation types according to the dimensions of flexibility and deregulation (Fig. 2.5). Compared with other types of cooperation, such as strategic alliances, company networks, or virtual enterprises, according to Freitag, Kück, and Becker (2016), resource sharing is the most flexible type of cooperation. Additionally, resource sharing is highly deregulated, thus implying low contractual complexity and legal regulation. The authors have described two dominant models of sharing resources: first, the usage of temporal overcapacity offered to other companies, and second, the joint investment into specific resources that would either be too expensive or not be fully utilised by single companies. In terms of sharing truck resources, both models are feasible. At the actor level, from the status quo perspective of an individual fleet operator holding its own assets, reducing temporary overcapacity or bridging undercapacity is possible by using shared vehicles. At the network level, pooling the vehicles of multiple players in a community open to all community members would correspond to model two.

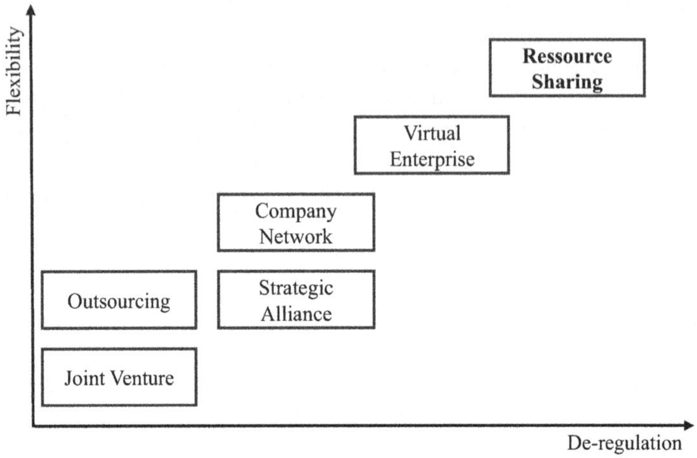

Fig. 2.5 Two-dimensional classification of cooperation types[6]

The depth and extent of horizontal cooperation can vary. Lambert, Emmelhainz, and Gardner (1999) have differentiated three types of cooperation depending on the level of integration, on a continuum from 'cooperation on demand' to 'horizontal integration' (Fig. 2.6). Performing more tasks and phases of the exercise jointly is associated with a higher degree of integration:

- *Type 1:* This type of cooperation is characterised by its short-term nature, typically focusing on the coordination and execution of joint activities. In this arrangement, only specific task areas or departments are involved to a limited extent.
- *Type 2:* Type 2 partnerships extend beyond the mere coordination and joint implementation of activities, and involve the integration of more comprehensive phases of execution, including joint control and planning of multiple logistical tasks. In type 2, compared with type 1, a greater number of company divisions and functions are engaged. Although not designed to be permanent, this type of partnership generally has a long-term orientation.

[6] Freitag, Kück and Becker (2016).

- *Type 3:* Type 3 cooperation represents an even higher level of integration, wherein all logistical sub-areas and phases of operations are conducted collaboratively. The level of integration is so extensive that the involved parties perceive their cooperation partners as extensions of their own companies.

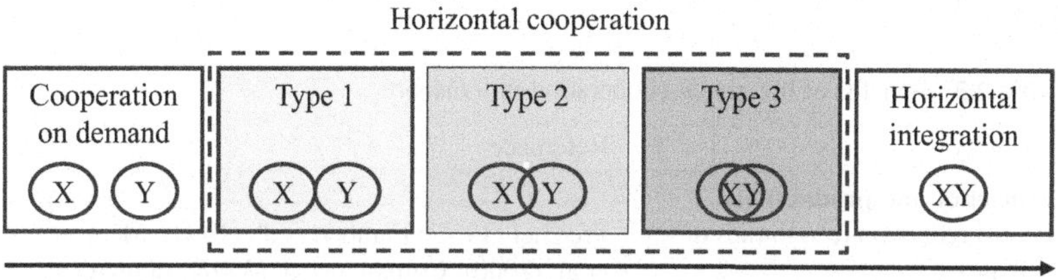

Fig. 2.6 Levels of integration in horizontal cooperation[7]

As shippers usually outsource their transport activities to LSPs to focus on core competencies, optimisation of transport efficiency is driven primarily by LSPs and their logistics networks. Thus, TS is usually investigated from the perspective of LPSs focusing on horizontal cooperation with other LSPs. Shipper and carrier cooperation is usually investigated separately, because of differing perspectives, objectives, and benefits (Chabot, Bouchard, Legault-Michaud, Renaud & Coelho, 2018). In this thesis, the focus is on carrier cooperation.

The following Sect. 2.2.2 outlines the objectives, and the drivers and barriers to sharing and transfers these to TS.

2.2.2 Benefits of Horizontal Cooperation

The initial driver of forming or entering into cooperation is the expectation of an overall net positive outcome for the cooperation project and the stakeholders involved (Parkhe, 1993). Benefits might be leveraged only by joining forces, and synergies might not be realised by either company acting alone. Bartlett and Goshal (2004) have identified three approaches to achieve benefits: pooling resources; sharing and leveraging specific capabilities and strengths of companies; and mutually exchanging complementary resources to achieve overall benefits.

[7] Based on Cruijssen (2006).

Horizontal logistics cooperations have been extensively studied and are a critical factor in the performance of LSPs (Caputo & Mininno, 1996; Schmoltzi & Wallenburg, 2012; Naesens, Gelders & Pintelon, 2009). Because of the broad field of application, benefits of horizontal cooperations address multiple categories, as indicated by the literature derived findings in Table 2.3. Major dimensions of benefits found in literature are (1) efficiency and productivity, (2) service level, (3) competitiveness, and increasingly (4) sustainability.

Table 2.3 Benefits of horizontal cooperation in transport

Category	Reference
Efficiency and productivity	
Cost reduction (economies of scale)	Freichel (1992); Lambert et al. (1999); Mentzer et al. (2000); Cruijssen et al. (2007); Ergun et al. (2007); Montoya-Torres et al. (2016); Chabot et al. (2018)
Operational efficiency	Cruijssen et al. (2007); Verdonck et al. (2013)
Resource utilisation	Liu et al. (2010); Dai et al. (2012)
Skilled workforce	Mentzer et al. (2000); Zineldin (2004)
Increased flexibility	Verdonck et al. (2013); Lehoux et al. (2014)
Service level	
Specialisation	Mentzer et al. (2000); Cruijssen (2006)
Service level improvement	Freichel (1992); Chabot et al. (2018); Cruijssen et al. (2007); Schmoltzi et al. (2012)
Transport frequency	Agarwal & Ergun (2010)
Competitiveness	
Market access/penetration	Freichel (1992); Agarwal & Ergun (2010); Schmoltzi et al. (2012); Cleophas et al. (2019)
New products/services	Lambert et al. (1999); Zineldin (2004)
Protect market share	Agarwal & Ergun (2010); Verdonck et al. (2013)
Shorter lead times	Schmoltzi et al. (2012); Singh et al. (2018)
Sustainability	
Less carbon emissions	Ballot et al. (2010); Pan et al. (2013); Karam et al. (2021)
Less distance travelled	Wang et al. (2015)
Less empty running	Ergun et al. (2007); Cruijssen et al. (2010)
Less traffic congestion	Montoya-Torres et al. (2016); Gansterer et al. (2018)

The primary motivations for engaging in horizontal logistics cooperation are improving cost efficiency and productivity, increasing customer service levels, and enhancing market position. Joint planning and the shared use of skilled employees can lead to significant cost reductions and productivity gains. By pooling resources, companies can achieve economies of scale, thereby lowering operational costs. This type of cooperation can lead to improved customer service through the provision of complementary services, adherence to higher service level agreements, and increased specialisation. The cooperative approach enables companies to offer more comprehensive and reliable services to their clients. By penetrating new markets, serving larger clients, protecting market share, and accelerating market entry, horizontal cooperation helps companies strengthen their market positions. Collaborative efforts enable firms to access markets that might have been out of reach to individual firms and to respond more quickly to market demands. Collaboration among LSPs not only enhances economic performance but also contributes to environmental sustainability. By reducing the number of vehicles required for transportation through shared logistics resources, companies can lower their environmental footprints. This decreased vehicle use translates into decreased emissions and less congestion, thus contributing to broader environmental goals.

In a literature review on sustainable collaborative transport, Aloui, Hamani, Derrouiche, and Delahoche (2021) have found that routing is the main problem faced by cooperations that can be addressed by horizontal cooperation followed by profit sharing, network design, distribution organisation, and inventory. This finding is reflected in the distribution of publications on the three pillars of sustainability: predominantly economic objectives are addressed, and some publications also include environmental considerations. However, the social dimension has not yet been analysed in relation to cooperation.

Although the formation of horizontal cooperations is typically unplanned and arises from personal relationships among LSPs (McKinnon, 2018), Sternberg, Linan, Prockl, and Norrman (2021) have observed increasing involvement of facilitators from outside the supply chain becoming sponsors of horizontal cooperation initiatives. Outside facilitators are considered relevant enablers of identifying potential members and opportunities and overcoming barriers. In reality, many cooperative initiatives fail and do not achieve the desired benefits. In the existing literature, drivers and barriers to horizontal cooperations have been identified to determine the success of initiatives (Cruijssen, Dullaert & Fleuren, 2007). Understanding the barriers to, and drivers of, effective cooperation can help LSPs maximise the benefits of their horizontal cooperation.

2.2.3 Challenges in Horizontal Cooperation

A major obstacle to horizontal collaboration is the existence of contrasting interests among participating organisations. As highlighted by Cruijssen et al. (2007), the involvement of multiple companies in a cooperation may give rise to situations in which the respective business objectives, operational strategies, and priorities of each company are distinct, thus impeding the alignment of goals and the achievement of mutual benefits. Such misalignment can result in conflicts and decreased overall effectiveness of the cooperation. Furthermore, the sharing of sensitive information and resources is often hindered by a lack of trust, which can result from differing agendas (Pomponi, Fratocchi & Rossi Tafuri, 2015). Partners might be reluctant to fully commit to a cooperative arrangement without assurance that their interests will be protected.

Another major challenge to effective horizontal cooperation is the high cost of coordination. The coordination of schedules, routes, and load capacities among multiple companies requires the implementation of sophisticated planning and the incorporation of real-time adjustments, which can be both time-consuming and costly. Verdonck et al. (2013) have discussed the substantial investment required in information and communication technology (ICT) systems to facilitate such coordination. The implementation of these systems poses a substantial financial burden, particularly for smaller companies, because of the needs for dynamic route planning, load matching, and real-time tracking. Furthermore, the ongoing expenses related to system upgrades and technical support exacerbate this challenge.

The costs of identifying and establishing relationships with reliable partners pose an obstacle to horizontal cooperation. Locating potential partners, evaluating their reliability, and negotiating terms are inherently time-consuming and resource intensive. The establishment of trust with new partners is a lengthy and complex undertaking that frequently necessitates a history of interactions and a demonstrated track record of reliability. As observed by Wang and Kopfer (2014), these search costs can impede the initiation of cooperative endeavours. Sternberg et al. (2021) have identified the role of facilitators from outside the supply chain as favourable and sometimes even necessary to establish horizontal cooperation beyond dyadic relationships.

The equitable distribution of gains derived from horizontal cooperation poses a substantial challenge. The assurance that all partners receive a fair share of the benefits, such as cost savings and increased profits, can be challenging, particularly when the contributions of each partner are not readily quantifiable. Cruijssen et al. (2007) have highlighted the need to establish a transparent and mutually agreed-upon framework for the distribution of gains and workloads, to prevent conflicts and ensure long-term satisfaction among partners. Inequities in distribution may potentially give rise to feelings of dissatisfaction and erode the foundations of the cooperative arrangement.

Disagreements and conflicts inevitably arise in any cooperative arrangement. Therefore, having effective negotiation and conflict resolution mechanisms in place is essential. In the context of TS, disagreements may arise regarding operational decisions, the allocation of resources, or the interpretation of agreements. Establishing transparent and fair decision-making processes is of the utmost importance in effectively managing and resolving any disagreements that may arise. As observed by Verdonck et al. (2013), the absence of robust conflict resolution mechanisms can prolong disputes, and consequently erode trust and cooperation among partners.

The indispensable ICT costs associated with horizontal cooperation in TS cannot be overstated. The deployment of advanced ICT systems is essential to facilitate real-time communication, data sharing, and coordination among partners. These systems facilitate dynamic route planning, load matching, and tracking, which are crucial for optimising the efficiency of shared logistics operations. Ballot and Fontane (2010) have emphasised the substantial financial commitment required for these technologies, which can be a major obstacle, particularly for companies with limited resources. The recurring costs associated with maintaining and upgrading ICT infrastructure further intensify this challenge.

The potential benefits of horizontal cooperation in TS can be concluded to be substantial; however, the barriers to successful implementation of cooperation are equally significant. Substantial obstacles to the implementation of horizontal cooperation include divergent interests among partners; high coordination and ICT costs; challenges in finding and trusting new partners; and difficulties in equitably distributing gains and workload, and in effectively managing disagreements. To address these barriers, strategies must include building trust, investing in technology, and implementing fair and transparent cooperation practices. By overcoming these challenges, companies can unlock the full potential of horizontal cooperation, thus achieving more efficient, cost-effective, and sustainable operations.

Given the current limitations of a cost-only focus for LSPs, the need to integrate cost efficiency and service levels with sustainability considerations is increasingly recognised. This integration is being pursued through cooperative initiatives with other LSPs. In the context of transport activities, the focus of horizontal cooperation is on capacity utilisation rather than on the direct exchange of assets. Despite the substantial benefits realised in practice, scientific studies have indicated room for improvement in truck utilisation. The sharing economy is a further stream of activity aimed at addressing this challenge.

Bridging horizontal cooperations and the sharing economy
Cooperations can be successful in the long term only if the associated benefits for individual participants outweigh the costs. This aspect applies to both horizontal cooperations and sharing concepts. The sharing economy and horizontal cooperation are both aimed at improving resource utilisation, decreasing costs, and minimising environmental impacts through shared use of assets and cooperative efforts. However, differences exist in their scope, participants, asset ownership models, and operational mechanisms.

2.3 Sharing Economy

In recent years, the sharing economy has received substantial attention from society and academia as sharing models are increasingly adopted. Companies such as Airbnb and Uber are well known for unlocking the potential of shared services in accommodation and transport through new business opportunities. The sharing economy is known by various names and substantially overlaps with other concepts such as the collaborative economy, collaborative consumption, access economy, platform economy, and community-based economy (Belk, 2014). Sharing models typically focus on underutilised assets to improve efficiency and sustainability (Hossain, 2020). Observed phenomena of shared goods and services involve swapping, trading, or renting as core activities offering asset owners higher utilisation rates (Hamari, Sjöklint & Ukkonen, 2016). Because of the wide field of application, the sharing economy has been defined in many ways, in both broad and narrow senses (Acquier, Daudigeos & Pinkse, 2017). Muñoz and Cohen (2017) have emphasised the relevance of efficiency defining sharing economy as

> "*a socio-economic system enabling an intermediated set of exchanges of goods and services between individuals and organizations which aim to increase efficiency and optimization of sub-utilized resources in society*" (p. 21).

This definition is well suited to the context of efficient freight transport through TS and is adopted to support understanding of the sharing economy in this thesis.

The presence of digital platforms is a defining characteristic of the sharing economy facilitating these exchanges (Hamari, Sjöklint, & Ukkonen, 2016). The sharing economy involves primarily three types of actors: service providers, users, and platform providers responsible for platform infrastructure and handling transactions (Benoit, Baker, Bolton, Gruber & Kandampully, 2017). Transactions typically do not result in changes in ownership. Sharing economy solutions foster cooperation between users and service providers, as both sides gain access to new partners. Horizontal cooperation, as outlined in Sect. 2.2.1, is an important component of many sharing concepts (Pomponi, Fratocchi & Rossi Tafuri, 2015). Motives and benefits for engaging in the sharing economy differ according to roles.

Benefits of the sharing economy for actors
The success of sharing initiatives lies in bundling complementary interests, thus creating advantages for the players involved on both sides of the market. According to Carissimi and Creazza (2022), on the demand side, users benefit from access to resources with immediate and flexible access, new partnerships, and reduced search costs. The final two points are also relevant to the provider side. In addition, providers benefit from additional income for the period in which the underutilised resources are used. As a third actor, platform providers typically pursue monetary interests and profit from transaction fees. In addition, successful implementation results in overarching positive effects at a social level, such as decreased greenhouse gas emissions and traffic congestion due to better vehicle utilisation.

2.3.1 Shareable Resources in Logistics

Whereas various business models related to sharing have been established according to Hossain (2020), transport and logistics are an application that can benefit from the SE. related to passenger transport, Nansubuga and Kowalkowski (2021) have found that car sharing has received high attention over the last decade, thus leading to the development of new business models and partnerships of manufacturers with other organisations (Hossain, 2020). However, the field of application is not limited to passenger transport and also includes freight transport. Facilitated by digital technologies, new market participants have entered the freight transport market with digitally enabled business models. Sennder is a prominent example of a digital freight forwarding company operating a digital platform that connects shippers with carriers. The platform serves as a marketplace where truck supply meets the demand for freight transport. This marketplace exemplifies a form of sharing capacities in the road freight market. In general, the area covered by shared use is broadly diversified in logistics.

Shareable resources in logistics
In classifying the types of resources shared, physical resources can be distinguished from intangible resources (Daudi, 2018). Physical resources include warehouses, transport vehicles, or infrastructure, whereas shared non-physical resources typically include information or knowledge sharing. Moreover, human resources form another dimension of resources; e.g., workforce sharing is increasing in project and freelance work environments. In 2017, Ocicka and Wieteska (2017) identified transport and warehousing the most important fields of sharing in logistics. The authors have highlighted that unlocking sharing potential is not only driven by cooperating companies but also involves potential competitors, thus describing the state of coopetition, cooperation in a specific field while competing in the market. This observation is in line with the competitive market situation outlined in Sect. 1.1, indicating increased pressure for LSPs to identify synergies with other market participants.

2.3.2 Differentiating Capacity and Asset Sharing

Regarding the types of resources shared, capacity sharing and asset sharing are two distinct concepts. However, the terms are rarely differentiated explicitly. Although trucks are shared, the nature of sharing may differ. On the one hand, an entire truck might be lent. This scenario is equivalent to truck rental, such that the vehicle is temporarily used by another person or company for the duration of the rental. The term 'asset sharing' is also used for this scenario. On the other hand, the load space of a truck alone can be shared without the truck changing users. This type of cooperation is typically found in freight exchange contexts, wherein loads and loading spaces are offered via a digital platform. The term used to describe such sharing approaches is 'capacity sharing'. On the basis of this background, both sharing concepts are briefly introduced and terminologically differentiated regarding trucks.

Capacity sharing: Capacity sharing involves utilising the unused or underutilised space in trucks owned by different companies to transport goods. It does not necessarily involve sharing the physical asset (the truck) itself but instead involves optimising the available space within a truck to carry additional cargo from other shippers. The motivations for capacity sharing include higher utilisation rates of empty load space and route optimisation to reduce transport costs.

Asset sharing: Asset sharing involves the joint use of physical assets, such as trucks, among multiple companies. This process often involves sharing the ownership or rental of the truck itself, rather than just the load space within it. In the case of permanent solutions, ownership can be shared, and the capital expenditure for a single company can be reduced. If the sharing is temporary, the motivations are balancing fluctuations; increasing flexibility by spreading the costs and responsibilities regarding the physical assets across multiple companies; and ensuring that trucks are in constant use.

Both models enhance logistics efficiency and reduce costs. However, they use different mechanisms to achieve these goals. Capacity sharing optimises the space within existing trucks, whereas asset sharing optimises the use of the trucks themselves. On the basis of this distinction, the conceptual understanding of TS in Chap. 4 follows the idea of an asset sharing concept.

2.3.3 Framework Conditions of Resource Sharing Concepts

Despite the many details that differ among sharing types, fundamental favourable and unfavourable framework conditions for resource sharing have been identified. A literature analysis of relevant conditions and influencing factors revealed central conditions favouring or hindering resource sharing (Table 2.4).

The success of a sharing solution depends on multiple variables, including technological, personal, and organisational factors. Because some of these factors can be both favourable and unfavourable, the specific design of a concept is a decisive factor in success or failure. Digital platforms fulfil a core task for the functioning of sharing models in practice, fundamentally bringing together supply and demand, thereby assuming a coordinating role.

Depending on the design of the mechanism, user transactions are associated with more or less effort, thus directly influencing the perceived cost-benefit ratio.

The implementation of a TS concept for road freight transport is compatible with the use of a digital platform and therefore includes a technological component. The market penetration of a technology requires user acceptance. The following section describes the link between sharing and technology acceptance, by using the example of TS.

Table 2.4 Overview of conditions favouring and hindering resource sharing

Category	Favouring conditions	Hindering conditions	References
Platform/ technology	Intelligent; neutral; decision support; actor-controlled	Employees' lack of knowledge regarding innovative logistics concepts	Cruijssen et al. (2007); Verstrepen et al. (2009); Karam et al. (2021); Mirzabeiki & Aitken (2022)
Organisation/ coordination	Preparedness for organisational change; joint decision-making; dynamic, centralised management; control/ monitoring system; continuous evaluation; needs-based distribution of resources; clear separation of cooperation and competition areas	Cooperation between organisational areas; changes in the corporate environment; prioritisation of distribution; time pressure and strict specifications; loss of control	Zineldin (2004) Verstrepen, et al. (2009) Islam & Olsen (2014) Daudi (2018) Sternberg et al. (2021) Mirzabeiki & Aitken (2022)
Cost-benefit ratio	Overall positive; holistic; low initial costs for each individual; fair cost and profit distribution system	High transaction costs; uncertainty regarding economic and environmental benefits	Cruijssen et al. (2007) Sternberg et al. (2021) Mirzabeiki & Aitken (2022)
Information sharing	Transparency; openness	Missing or unilateral information flow; inaccuracy; insecurity	Karam et al. (2021)
Trust	High degree	Mistrust; prejudices, competitiveness	Cruijssen et al. (2007) Pomponi et al. (2015) Daudi (2018) Karam et al. (2021)
Security	Planning robustness; protection of sensitive data	Data protection risks; lack of security	Daudi (2018)

(continued)

Table 2.4 (continued)

Category	Favouring conditions	Hindering conditions	References
Contractual	Fair; clear; easy access; low complexity particularly initially	Drafting a suitable contract; disagreement regarding the form of cooperation	Cruijssen et al. (2007) Verstrepen et al. (2009) Karam et al. (2021) Sternberg et al. (2021)
Partners	Mature; reliable; similar market power; similar corporate structure/ strategy; trust; converging interests; complementary resources	Difficulties in identifying proper partners; diverse customer requirements; different strategies and visions, equal bargaining position	Mentzer et al. (2000) Cruijssen, et al. (2007) Verstrepen, et al. (2009) Karam et al. (2021) Makaci et al. (2017) Melander & Wallström (2022)

2.4 Truck-Sharing

Research on TS has not yet been established as an independent scientific field. TS being focused on a specific use case of the sharing economy, relevant findings are found in several different literature streams. Drawing on industrial logic, research on horizontal cooperation provides insights into the prerequisites, opportunities, and risks that can be transferred to TS. From an overarching industry-independent perspective not restricted to freight transport, research on the sharing economy provides impulses for TS as outlined in the previous Sect. 2.3.

In recent years, several studies have explicitly referred to the concept of TS, but a clear definition and explanation of what is meant by this term is lacking. The aim of Sect. 2.4.1 is to categorise this literature to create a basis for a conceptual foundation in Sect. 2.4.2, to provide an understanding of this work. In sects. 2.4.3 and 2.4.4, the concepts of psychological ownership (PO) and technology acceptance are addressed and are related to TS, as relevant to the context of the thesis.

2.4.1 Truck-Sharing in Light of Established Research Streams

To date, the research literature on TS in road freight transport is limited in scope and content. This field of study has been advanced by Islam and colleagues (Islam & Olsen, 2014; Islam, 2018; Islam, Shi, Ahmed & Uddin, 2019; Islam, Uddin, Shi, Sharif & Ahmed, 2021), who have stated that the pursued core idea of TS is "*to increase the overall loading of a truck by completely reducing fully empty or partially loaded miles*" (Islam et al., 2019, p. 642). That research group has provided insights into challenges for TS specifically related to hinterland container transport (Islam & Olsen, 2014), challenges (Islam et al., 2019) and underlying motives from the shipper side to initiate sharing initiatives.

The objectives targeted by TS correspond to those of horizontal cooperation and are aimed at the inefficiency identified in Sect. 2.1. Islam and Olsen (2014) have postulated the following benefits of the sharing concept in the context of transport: an increase in the utilisation rate of vehicles, an increase in earnings for truck drivers, a reduction in emissions and pollution, and resolution of traffic problems by reducing the overall number of truck trips. Islam et al. (2021) have expanded on the motivations for collaborative TS initiatives to include customer-centric dimensions, such as responsiveness of transport solutions, convenience-seeking behaviour, and security of transport operations.

In line with McKinnon (2000), Islam and Ohlsen (2014) consider empty trips to be the most significant form of truck underutilisation. This addressed inefficiency coincides with the category of primary efficiency measures illustrated in Fig. 2.3, covering both empty running and utilisation rate.

However, the research to date coining the term TS is limited by its focus on sharing transport capacities or transport services, but not on the exchange of the physical asset of the vehicle.[8] According to the idea of the sharing economy, the focus of TS is not optimising routes and truck utilisation, which is already being practiced with assistance from planning tools. Therefore, prerequisites and barriers to TS outlined in the scarce body of literature on TS are not further

[8] A terminological demarcation of asset and capacity sharing approaches is provided in Sect. 2.3.2.

discussed in detail herein, because they overlap with findings derived from the literature in sects. 2.2.2 and 2.2.3.

Asset sharing

An understanding of TS as an asset sharing concept is introduced by Stölzle and Wildhaber (2019). The two authors introduce the term TS with regard to the application in general cargo cooperatives. According to the authors "*Truck Sharing provides the possibility to share underutilized trucks (as a physical asset) that are owned by logistic service providers via a community-based (digital) platform with other logistics service providers*" (p. 77).

Drawing on the core aspect of asset sharing within the definition, this understanding emphasises the industry-wide approach between fleet operators and acknowledges the role of technology as an organisation enabler for matching processes between the supply and demand sides. However, in contrast to the above presented definition, in this thesis, the application of TS is approached in an open-ended manner and is not limited to general cargo cooperations. The reason for this decision is the disadvantage of closed communities such as general cargo cooperations, with as many as 250 companies involved. Compared to an open market approach addressing thousands of companies, closed-shop approaches would severely limit the number of users involved upfront, thus limiting potential network effects.

In a recent study, Mrabti, Hamani, and Delahoche (2022) have emphasised the significance of flexibility in enhancing all three dimensions of sustainability, in particular by reducing logistics costs, CO_2 emissions, noise levels, and accident risk. Consequently, greater flexibility might be posited to be beneficial by enabling exploitation of a greater number of alternatives and optimising potential solutions to improve transport operations. Increased flexibility can have a direct impact on vehicle downtime, which can be eliminated by renting out trucks at times when they were previously not in use. Primary efficiency measures might be positively influenced if a sharing partner finds a 'better' use case for the truck that increases truck utilisation, and/or decreases empty kilometres and the overall distance travelled.

By combining the outlined efficiency measures in Sect. 2.1 with the research streams of horizontal cooperation and the sharing economy in Sects. 2.2 and 2.3, the subsequent Sect. 2.4.2 outlines terminological and conceptual fundamentals of TS to establish a shared understanding of TS among researchers and practitioners, and to provide a frame of reference for investigation of TS in this thesis.

2.4.2 Conceptual Fundamentals of Truck-Sharing

Because a common understanding of the term 'truck-sharing' has not yet been established in the scientific community, as outlined in Chap. 2, and the term is therefore associated with different concepts, the following definition is provided as the conceptual basis of this thesis:

> *"Truck-sharing involves the shared use of trucks owned by different transport companies, with the trucks being physically exchanged between the companies involved."*

This definition emphasises the significance of the physical exchange of vehicles between companies as a key criterion. Simultaneously, the definition clearly delineates sharing capacity or load space, a form of cooperation that is currently widespread in various forms (e.g., by using freight exchange platforms) in practice and does not involve the exchange of vehicles. Instead, shipments are exchanged while the operating company of a truck remains unchanged.

TS implies flexible, demand-orientated exchange of vehicles between fleet operators. The physical exchange of assets is at the centre of the sharing concept. Underutilised truck resources are made available to other market participants. Depending on the needs of a company, the exchange can occur with or without the involvement of truck drivers. In contrast to other areas of application, such as car sharing, the sharing models to date have formed a niche in the trucking market and are at most applied bilaterally between two companies familiar to each other. TS enables temporarily unused trucks to be made available to other transport companies, thus allowing resources to be used more flexibly. This pooling of capacities facilitates the balancing of different peaks in demand at an aggregated level with a smaller total number of trucks.

As an innovative strategy designed to enhance the efficiency and flexibility of road freight transport, the conceptual framework of TS can be understood through its constitutive attributes. This conceptual advance has yet to be rooted in practice. The following concept should be considered an anticipatory conceptual frame of reference:

- *Provision of sharing opportunities*: TS enables fleet operators to offer and access sharing opportunities for underutilised trucks. This concept targets the efficient use of available truck resources, ensuring that trucks are not left idle when they could be used by other transport companies.

- *Physical asset utilisation*: At its core, TS focuses on the trucks themselves as physical assets. The goal is to optimise the use of these assets by sharing them among different providers, thereby increasing the utilisation rate of existing vehicles and reducing the overall number of trucks needed. As described in Sect. 2.3.2, in contrast to the focus of freight exchanges on capacity sharing, TS is considered an asset sharing concept.
- *Ownership by LSPs*: The trucks involved in TS schemes are typically owned by LSPs operating trucks themselves. This horizontal ownership structure allows providers to retain control over their assets while participating in a shared economy model in which they can lend their underutilised trucks to others in need.
- *Digital platform*: The sharing process is facilitated by a community-based digital platform organising the matching process. This platform acts as a marketplace where LSPs can list available trucks and find trucks that meet their immediate needs. The digital nature of the platform ensures real-time updates, ease of access, and efficient matchmaking between supply and demand.
- *Collaboration among LSPs*: TS promotes a collaborative approach among LSPs. By sharing resources, providers can support one another in meeting fluctuating demand, thereby improving the overall efficiency and resilience of the logistics network.

Carrier focus

Since shippers primarily focus on procuring transport services outsourcing transport operations to dedicated transport companies, shippers are most concerned with costs and service quality while carriers are responsible for providing optimised transport solutions at minimum costs. In this context, carriers are operationally responsible for transport execution including route planning, freight consolidation and fleet management. Efficiency increases through TS enabled transport optimisation are approached from the perspective of carriers.

Demarcation from truck rental

Because shippers focus primarily on procuring transport services and outsourcing transport operations to dedicated transport companies, shippers are most concerned with costs and service quality, whereas carriers are responsible for providing optimised transport solutions at minimum cost. In this context, carriers are operationally responsible for transport execution including route planning, freight consolidation, and fleet management. Efficiency increases through TS enabled transport optimisation are approached from the perspective of carriers.

Distinction with respect to truck rental

In the context of the sharing economy, the terms 'sharing' and 'renting' overlap, because sharing concepts involve a form of renting when an owned asset is made available to another party. Because truck rental is well established to account for a substantial and rising share of one-quarter of the total truck market, according to Gnamm, Stricker, and Bendig (2010), TS is to be distinguished from truck rental.

Renting trucks enables transport companies to increase the flexibility of their fleet by increasing the proportion of trucks rented. Renting is often considered in terms of risk minimisation to buffer fluctuating transport demand. In this respect, the objectives of sharing and renting are identical.

Supply in the truck rental market is provided by a variety of companies that do not directly operate as transport companies themselves, such as manufacturer-independent truck rental companies and truck manufacturers as rental companies. Thus, a clear distinction exists between sharing and the classic rental market, according to the role of the parties providing the rental offer. Whereas TS pursues a horizontal peer-to-peer approach to network fleet operators with one another, cooperation in the traditional rental market occurs vertically with companies at different levels of the value chain (Fig. 2.7).

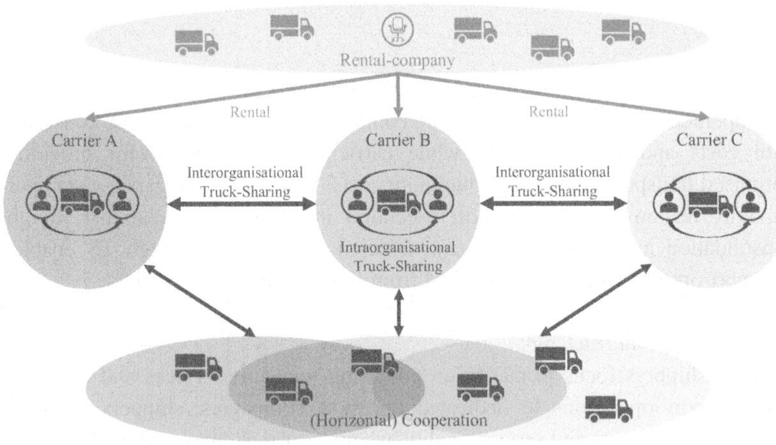

Fig. 2.7 Distinction between truck-sharing forms and truck rental

Interorganisational and intraorganisational sharing

The horizontal sharing approach is applicable in several ways. Such sharing can occur as inter-organisational TS across company boundaries or intraorganisational TS within a company. The second approach results from the industrial logic in which, in transport networks of large transport companies with many depots, the individual depots often act autonomously and strive for self-optimisation, because of their own economic responsibility. Under regionally fluctuating freight transport volumes, the shifting of assets between depots primarily provides an opportunity for intraorganisational sharing. In the context of this thesis, however, the focus is on interorganisational sharing, because the challenges of intraorganisational cooperation are comparable and can be derived. One variant of interorganisational sharing is joining forces in a cooperation, an arrangement synonymous with restriction of the group of stakeholders. As shown in Sect. 2.2, cooperation is widespread in transport networks; therefore, cross-company cooperation can be built on existing cooperations. However, the smaller the community of fleet operators, the smaller the potential network effects. Sharing is therefore not limited to closed communities in the context of this thesis.

Rental duration

A distinction can be made between short-term and long-term rentals on the classic rental market. Whereas short-term rental spans days to weeks, often to bridge the period of maintenance work, long-term rental spans 6 months to approximately 3 years. Because the administrative costs of truck rental are incurred regardless of the rental period, long-term rentals from several months onwards are a particularly attractive business case from the rental company's perspective. Short-term rentals are offered to customers for service reasons but are generally not profitable for rental companies and therefore are not strategically promoted, as confirmed through preliminary discussions with rental companies and LSPs. In principle, TS is feasible over short and long periods. Because of the already broad coverage of long-term rental by the classic rental market, shorter rental periods of several or single days are pronounced. TS is therefore aimed at tapping into the ultra-short-term rental segment.

Truck or trucker sharing?

Transporting goods by road requires a truck, with either a dedicated load space or a separate trailer, and a truck driver to be operational simultaneously. Each of these three components—vehicle, load space, and truck driver—is therefore a limiting factor (Fig. 2.8).

Fig. 2.8 Vehicle, load space, and truck driver as prerequisites for the provision of road freight transport services

Given the growing truck driver shortage in Europe, as outlined in Sect. 1.1, truck driver availability is increasingly becoming a bottleneck hindering the success of sharing initiatives. To be utilised by the market, TS must be accepted by key stakeholders including truck drivers. Thus, beyond convincing transport companies, a TS concept must also address truck drivers' perspectives, including the organisation of the sharing process from the driver's perspective and the question of what is shared. TS involves potentially sharing solely trucks, or both trucks and truck drivers, thus leading to implications including labour law dimensions, as discussed in Chap. 5.

Organising the sharing process

The utilisation of a digital platform is fundamental to the organisation of TS, as it facilitates the complex coordination of supply and demand. In a dynamic environment such as road freight transport, an immense quantity of information is generated, including details regarding available vehicles, routes, timetables, and specific freight requirements. An algorithmic platform is capable of processing this data in real time, enabling the identification of suitable matches with greater expediency than manual processes. This would be challenging to achieve without a digital solution.

Regarding the organisation of TS, various operator models are possible, as described in Sect. 2.3. Whereas the bundling of truck supply and demand is a core function and can be organised via a digital platform, participant trust is a prerequisite for engaging both sides of the market and is influenced by the operator's role. Various approaches are conceivable (Fig. 2.9). Centralised coordination by an individual company with a connection to the transport industry is one option. Experience with digital marketplaces in logistics has indicated that other (competing) market participants have reservations regarding such an operator model, because independence is questioned. A second option is the cooperation of LSPs, as practised in existing logistics cooperations. The openness of such a platform to external transport companies remains to be addressed. Option 3 is operation by an independent third party from outside the industry. In this case, acceptance poses a low risk. However, a third party is likely to be profit-oriented and, this option, in contrast to a cooperative solution, would generate profits.

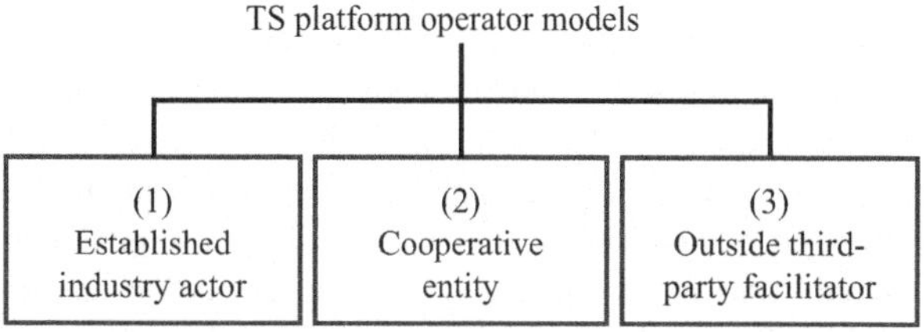

Fig. 2.9 Operator models for a truck-sharing platform

Potential designs for TS are illustrated in Fig. 2.10.

<table>
<tr><th>Parameter</th><th colspan="12">Configuration</th></tr>
<tr><td>Resources shared</td><td colspan="4">Truck</td><td colspan="4">Trailer</td><td colspan="4">Driver</td></tr>
<tr><td>Ownership</td><td colspan="6">Unilateral</td><td colspan="6">Shared</td></tr>
<tr><td>Cooperation type</td><td colspan="4">Horizontal</td><td colspan="4">Vertical</td><td colspan="4">Lateral</td></tr>
<tr><td>Community</td><td colspan="6">Open</td><td colspan="6">Closed shop</td></tr>
<tr><td>Geographical scope</td><td colspan="3">Local</td><td colspan="3">Regional</td><td colspan="3">National</td><td colspan="3">International</td></tr>
<tr><td>Location</td><td colspan="6">Station based</td><td colspan="6">Free float</td></tr>
<tr><td>Organisation</td><td colspan="4">Industry actor</td><td colspan="4">Cooperative entity</td><td colspan="4">Third party</td></tr>
<tr><td>Pricing</td><td colspan="4">Static</td><td colspan="4">Dynamic</td><td colspan="4">Differential</td></tr>
</table>

Fig. 2.10 Morphological box for truck-sharing design

Supplementing established efficiency levers

Whereas established efficiency levers such as route optimisation and cooperations have been instrumental in enhancing logistics efficiency, they have focused primarily on managing fixed fleets optimising route planning in the sense of maximised truck capacity utilisation at minimised empty kilometres with the shortest possible distance travelled. Route optimisation remains a core competency in freight dispatching and continues to offer substantial potential for efficiency gains. However, it does not specifically address the issue of idle times.

TS addresses this limitation, and the underutilisation challenge can be targeted directly. Instead of merely optimising routes and capacity utilisation, TS starts before operationally deploying trucks, thereby identifying the best use case for a truck through dynamic allocation facilitated by a digital platform. This approach not only maximises the use of existing assets but also introduces a layer of flexibility allowing LSPs to adapt to changing demands more quickly.

In essence, TS is a forward-thinking concept that supplements existing efficiency measures: it leverages collaborative practices and digital technologies to create a more adaptive and resource-efficient logistics ecosystem. By enabling logistic service providers to share their underutilised trucks, TS helps reduce idle times, lower operational costs, and improve overall sustainability in road freight transport.

Concluding normative vision

A future vision of TS in road freight transport might involve sharing trucks across company borders enabled by an open digital platform. Sharing would not be limited to peer-to-peer sharing but would be open to rental companies extending offers. The number of supply and demand requests on a platform involving thousands of companies would be so high that the next available truck would always be within several kilometres away, thus allowing LSPs to flexibly rent assets. The configuration of the platform would enable offers with or without a truck driver.

The potential adoption of this concept in the context of contemporary business operations gives rise to the conclusion that the emergence of TS would result in substantial modifications for both truck drivers and transport companies. Sharing, or using instead of owning, is a new industry concept whose operation requires technology. According to psychological research, people tend to prefer familiar and controllable conditions and often perceive change as a threat to their comfort zone and autonomy. The underlying concept of resistance to change dates to Lewin (1947). His three-phase model of change describes the process of change in organisations and social systems in three phases. In phase one, ‘unfreeze‘, breaking existing structures and habits is crucial. As applied to truck drivers and entrepreneurs, this process involves developing awareness that a change towards TS is beneficial. If this prerequisite is not met, change might fail in this first phase. The psychological construct of PO in Sect. 2.4.3 and the *technology acceptance model* from information systems theory in Sect. 2.4.4 introduce two concepts relevant to the TS context in this thesis.

2.4.3 Truck-Sharing and Psychological Ownership

Truck drivers spend most of their working hours in a vehicle. The driver's cab serves two distinct purposes as a place of work and a place of privacy where personal belongings can be stored. Truck drivers engaged in international long-distance transport, who are on the road for extended periods of time, often spend the night in the truck and therefore are closely attached to their vehicles. Although trucks are typically owned by transport companies, the level of truck driver attachment to vehicles is relatively high, as reflected by increasing truck customisation and incorporation of special equipment according to truck driver preferences (Sect. 3.4.3). Combined with established truck deployment profiles, wherein the truck is often used by only one driver, truck drivers develop feelings of using their own vehicles. Thus, a question arises regarding truck driver acceptance of TS, because, in contrast to individual use, TS is associated with a different utilisation model in which a truck is used by several users.

Possession and territorial behaviour

In this context, attention is drawn to the link between possession and territorial behaviour. Research has suggested that individuals can develop possessive feelings towards objects, regardless of legal ownership (Dawkins, Tian, Newman & Martin, 2015). Possession can cause people to protect and defend their perceived possessions. Accordingly, people who feel ownership of something are more likely to protect their possessions and even control and limit access to them (Wilpert, 1991). Such behaviour is referred to as *territorial behaviour,* defined by Brown, Lawrence, and Robinson (2005) as "*an individual's behavioural expression of his or her feelings of ownership toward a physical or social object*" (p. 2). Feelings of ownership may increase territorial behaviour, which in turn may increase and reinforce feelings of ownership. Given that in TS contexts, assets are exchanged—in contrast to established horizontal collaborations that focus on sharing capacity—the influence of ownership becomes relevant in this context. A market offer for vehicle sharing presupposes that asset owners are willing to grant temporary access to other actors. Ownership must be considered in investigating the willingness to participate in TS environments. The concept of PO can feasibly address this issue.

Psychological ownership
Pierce, Kostova, and Dirks (2003) have provided a definition for PO as a state in which an individual perceives ownership over an object as if it were theirs. PO is independent of actual or legal ownership. Paundra, Rook, van Dalen, and Ketter (2017) have investigated the role PO in car sharing services and have found that people with a high PO prefer owning a car to sharing one. They have concluded that considering PO is relevant in developing measures to stimulate car sharing services. One aspect of cars is the specific emotional attachment of individuals. Lord, Després, and Ramadier (2011) have found that people often have an emotional attachment to owning a private car, which is preferred over sharing services; this finding might be explained by people considering possessions as part of their self-identities. Findings from car-sharing usage behaviour are a suitable basis for comparison. Given the observed correlation between PO and territorial behaviour (Brown, Crossley & Robinson, 2014; Zhang, Liu, Zhang, Xu & Cheung, 2021), exploring PO is relevant to this thesis. I hypothesised that truck drivers considering a truck as a personal possession to a greater extent would be less open to TS (4.2.3). In this context, I investigated the extent to which the feeling of PO towards one's own truck develops, and which factors determine PO. Second, I investigated the transferability of findings from car sharing to how PO is related to willingness to share one's truck.

2.4.4 Truck-Sharing and Technology Acceptance

The success of sharing models is significantly influenced by perceived usefulness (PU). If sharing models generate added value for users and are useful overall, people are more likely to be open to using them. As shown in Sect. 2.3, digital platforms are indispensable enablers. As such, sharing models in general, and TS in particular, involve a technological component. Whereas TS should be perceived as fundamentally useful to gain acceptance, this also applies to the underlying technology.

The Technology Acceptance Model (TAM), a theoretical model developed to explain and predict how users accept and utilise new technologies, can serve as a reference model for technology acceptance (Davis, 1989). According to this model, PU and perceived ease of use (PEU) are predictors of intention to use

(IU) (Fig. 2.11). The TAM is particularly applicable to consumers' acceptance of information technology related products (Marangunić & Granić, 2015) and therefore was considered feasible in the context of this thesis, given that TS uses a digital platform to organise the matching processes. Liu and Yang (2018) have confirmed the applicability of the TAM in the sharing economy and have revealed that PU and PEU are the main factors affecting behavioural intention.

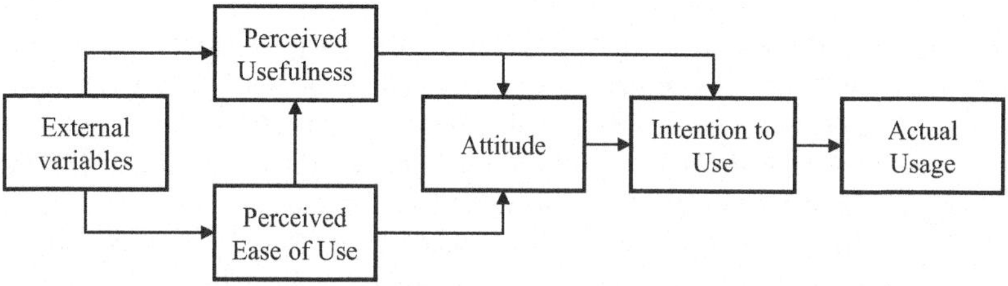

Fig. 2.11 Technology acceptance model[9]

The TAM builds on the theory of reasoned action developed by Fishbein and Ajzen (1975), aimed at explaining human behaviour bridging the relationship of attitudes and behaviours. The theory suggests that behaviour is determined by an individual's intention to perform a behaviour, which in turn is determined by attitudes and subjective norms. Building on this concept, further models were developed, such as the theory of planned behaviour by Ajzen; combined models; and, building directly on TAM, the unified theory of acceptance and use of technology. The last two models in particular are associated with a greater number of predictors and dependencies. With regard to enabling TS, a focus on behavioural intentions instead of actual usage is currently appropriate, given that the concept is not observable in practice.

To gain insights into drivers of, and barriers to TS, and design implications for TS implementation, one focus of this thesis is exploring favourable and unfavourable influencing factors. As outlined in Sect. 2.4.3, trucks are a high involvement object to which both truck drivers and owning companies attach emotional value similar to that of cars. The significance of PO regarding the acceptance of TS is therefore further examined in this thesis.

Conclusion: In the previous sections of Chap. 2, the concept of efficiency was initially introduced and expanded upon with reference to road freight transport. In addition, horizontal cooperation and the sharing economy were considered

[9] Based on Davis (1989).

the theoretical basis of the investigation. TS was embedded in this context, and a conceptual foundation was developed in Sect. 2.4.2 to foster a shared understanding of TS among researchers and practitioners. Sects. 2.4.3 and 2.4.4 introduced constructs relevant to investigating drivers and barriers in Sect. 4.2.

2.5 State of Research

Building on the previous sections of Chap. 2, this chapter outlines the current state of research. For this purpose, the background on the status quo of efficiency in road freight transport is linked to the theoretical background for horizontal cooperation and the sharing economy, on the one hand, and is merged with the conceptual fundamentals of TS, on the other hand.

The sharing economy is increasingly linked to transport and is considered a solution to challenges in road freight transport, particularly for addressing underutilised resources. The focus herein is on the potential to increase transport efficiency by augmenting capacity utilisation and reducing empty kilometres, efficiency being associated with both economic and ecological sustainability. Another approach to overcome these challenges is horizontal cooperation, which is widely used in practice and has been intensively investigated academically, according to a shared definition. Although various definitions of SE exist, a common understanding has also developed.

In contrast to bicycle- or car-sharing, a sharing model has not yet been established for heavy-duty traffic. Large-scale sharing of trucks cannot be found in commercial road freight transport. Approaches are typically restricted to dyadic solutions in a regional context among LSPs that are known to each other. Nonetheless, the term 'truck-sharing' has increasingly been used among researchers such as Islam and Olsen (2014), to refer to concepts of capacity sharing. The term is not used consistently, and a standardised definition is lacking. Stölzle and Wildhaber (2019) first explicitly described physical asset exchange as part of TS. On the basis of this background, conceptual fundamentals of TS are outlined in Sect. 2.4.2 to fundamentally address this shortcoming, according to the perspective adopted in the SE. Adopting an asset sharing perspective, this thesis is based on the understanding of Hamari, Sjöklint, and Ukkonen (2016) regarding collaborative consumption involving activities such as swapping, trading, and renting as part of TS.

Apart from the conceptual delimitation issues, all studies on the potential of TS theoretically address possible optimisations primarily from a top-down perspective. To date, there has been a lack of studies that focus directly on the freight operation of LSPs, current levels of efficiency and applied optimisation mechanisms as a basis for identifying specific use cases. There is also a lack of differentiated analyses with regard to the degree of utilisation, empty running and downtimes of trucks. The deployment profiles of trucks in road freight transport and the active market players are considerably diverse, meaning that a differentiated analysis with regard to segments, transport distances, industry requirements and truck types is essential in order to identify specific use cases.

Despite the many benefits of sharing, there is still a research gap regarding its applicability in a TS context. There is a lack of empirical observation and the question arises as to why sharing in the form of asset sharing has not yet been adopted in freight transport. Further, there is a research gap regarding barriers and the question of how to enable catalysts for adoption; these have been identified in general for SE but not specifically for TS.

Building on the current state of research and its shortcomings, guiding RQs are derived in Sect. 2.6.

2.6 Research Questions

In the context of current challenges in road freight transport, TS has emerged as a solution to enhance efficiency by addressing underutilised capacity and empty return trips. To provide a foundation for implementing TS, exploring specific fields of application that must be aligned with the business reality of transport companies is critical. To deliver added value for the market and individual market participants, use cases are measured according to the extent to which they increase transport efficiency:

RQ1: What **potential** and future **use cases** contributing to increased efficiency are emerging for TS in the context of road freight transport?

Whereas generic benefits of TS are documented, further investigation is required to gain insight into the factors that drive and hinder successful adoption of TS in road freight transport:

RQ2: What are the key **drivers** of, and **barriers** to, TS in road freight transport?

Unless due consideration is given to the implementation stage, the conceptualisation developed by TS will have limited value. To progress from the conceptual stage to practical implementation, actionable measures must be identified and implemented to facilitate successful adoption of TS. This process encompasses the development of technological, regulatory, and organisational strategies that can overcome existing barriers and catalyse the adoption of this concept. Addressing this research demand is expressed in the third RQ:

RQ3: What **implications** and **recommendations** can be drawn to leverage the adoption of TS in practice?

Research Approach Towards Truck-Sharing in Road Freight Transport

3

Chapter 3 outlines the research approach and the applied research methods, to build a foundation for the empirical access to this dissertation project.

3.1 Research Paradigm

To ensure the selection of an appropriate research design, methods, and data collection procedures, the paradigm underpinning this research is outlined to provide transparency. According to Saunders, Lewis, and Thornhill (2023), the alignment of a suitable research paradigm with the research design and methods is crucial for integrity and coherence.

The starting point of this dissertation arose from actual managerial challenges, aligning with Ulrich's (1984) understanding of business administration as a practical social science. This perspective focuses on the design and management of productive social systems. Ulrich has stated that social systems such as companies face daily management challenges in striving to find solutions to these problems.

This research follows an applied science perspective and is aimed at creating practically relevant and useful knowledge and solutions to critical practical challenges in road freight transport while enriching theoretical understanding (Dossabhoy & Berger, 2002). Specifically, this thesis focuses on the challenges and concepts associated with TS, to improve the efficiency of road freight transport operations.

L. Häberle, *Truck-Sharing in Road Freight Transport*, Supply Chain Management,
https://doi.org/10.1007/978-3-658-48754-6_3

3.2 Research Methodology

In studies in the logistics domain, researchers frequently apply theories from a range of disciplines, according to Defee, Williams, Randall & Thomas (2010). Given the novelty and limited research output on sharing related to trucks, little theoretical foundation existed to underpin the investigation of the role and potential of TS in road freight transport. In this emerging research field, studies have focused on exploration and discovery rather than applying specific theories. Therefore, an explorative research design is applicable to determine *"what is happening; to seek new insights; to ask questions and to assess phenomena in a new light"* (Robson, 2002, p. 59).

This exploration requires involving diverse stakeholders in practice. Therefore, this research followed an engaged scholarship approach to understand the research problem (Van de Ven, 2007).

3.2.1 Engaged Scholarship Approach

The engaged scholarship approach facilitates collaboration between academics and practitioners in understanding and resolving real-world problems, advancing knowledge, and providing relevant solutions to organisational challenges (Van de Ven, 2007; Van de Ven, 2018, Mathiassen, 2017). This method is particularly suitable to investigating complex, multifaceted problems, emerging practices, and innovation processes. Partnering with industry stakeholders enables gaining deeper insights and developing practical implications. The collaborative approach ensures that research outcomes are both academically robust and practically relevant, thereby effectively addressing challenges faced by organisations.

The value of engaged scholarship is well recognised to be applied in research fields that emphasise collaboration between academia and practice to address real-world problems. Wong (2021) has emphasised its importance in advancing theory and practice, and van Hoek (2021) has recognised its role in enhancing the managerial relevance of research. Contemporary challenges in these areas require innovative and impactful research, wherein engaged scholarship is a crucial driver of such change (Touboulic, McCarthy & Matthews, 2020).

Engaged scholarship focuses on interacting with various stakeholders to jointly generate knowledge and develop practice-oriented solutions. This approach is ideal for collecting diverse opinions, views, and perspectives, which can then be synthesised into comprehensive recommendations. Unlike direct intervention methods, engaged scholarship relies on the active involvement of stakeholders to ensure that the research remains relevant and grounded in a real-world context. By emphasising collaboration and leveraging diverse stakeholder insights, engaged scholarship is suited to advancing both academic knowledge and practical solutions to challenges related to sharing truck resources. From the four forms of engaged scholarship suggested by Van de Ven (2007), this thesis combines *collaborative basic research*, defined as sharing power and collaboration between researchers and stakeholders, and *design and evaluation research*, focusing on normative knowledge related to the design and evaluation of solutions to problems and specific recommendations.

For investigating the RQs, the research context is translated to the generic structure of engaged scholarship studies introduced by Mathiassen (2017), as illustrated in Fig. 3.1. The framework serves as a foundation to derive the empirical research design.

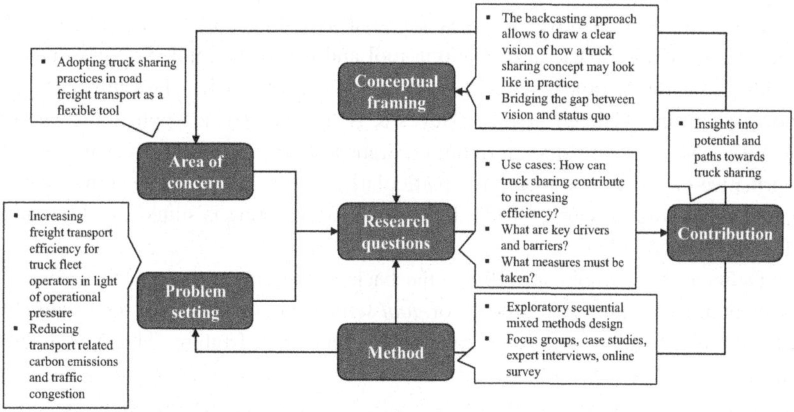

Fig. 3.1 Engaged scholarship approach to truck-sharing[1]

[1] Based on Mathiassen (2017).

3.2.2 Backcasting Methodology

The principal challenge in examining TS is the absence of a practical foundation. TS is neither empirically observable nor conceptually established, and thus a future-oriented conceptual framework must first be developed on the basis of the current challenges.

Research focusing on long-term change typically adopts future oriented approaches such as trend analysis, forecasting, scenario planning, and visioning scenarios. One further approach, developed in the 1970 s as an alternative to traditional energy forecasting, is backcasting. According to Robinson (2003), backcasting is defined as an approach to future studies involving *"the development of normative scenarios aimed at exploring the feasibility and implications of achieving certain desired endpoints, in contrast to forecasting studies aimed at providing the most likely projection of future conditions*" (p. 841). Instead of extrapolating future trends according to current data, backcasting centres on defining desired end-states and subsequently devising strategies to realise them (Robinson, 1990). Compared with other scenario approaches, backcasting is most suitable to exploring complex societal issues necessitating transformative change or paradigm shifts (Holmberg & Robert, 2000; Börjeson, Höjer, Dreborg, Ekvall & Finnveden, 2006). By focusing on desired end-goals, this process fosters innovative thinking and reveals novel avenues for goal attainment. In addition, backcasting is considered a backward planning tool and is widely applied across diverse fields to set future objectives and identify pathways to their achievement (Mendoza, Sharmina, Gallego-Schmid, Heyes & Azapagic, 2017). Application sectors include energy, chemistry, agriculture, transport, and tourism. Because of its normative character, the method is particularly well suited to sustainability applications (Vergragt & Quist, 2011). In addition, backcasting is suited to addressing TS, as outlined below.

Delineated by Robinson (1990), the backcasting approach comprises several sequential steps. First, the process of *goal setting* commences with the establishment of clear and specific objectives for the envisioned future. These goals are ideally visionary, challenging, and sustainable, and span various domains such as environmental stewardship, social equity, and economic development. Second, in the stage of *scenario development*, multiple scenarios are crafted to depict potential future states aligned with the established objectives. These scenarios are often crafted collaboratively, by integrating insights from diverse stakeholders to encompass a spectrum of perspectives and potentialities. Third, in *pathway development*, a detailed analysis is undertaken to delineate the requisite steps and interventions necessary to transition from the present to the desired future state.

This process entails identifying and evaluating pathways encompassing policy decisions, technological innovations, behavioural shifts, investments, and other pertinent factors. Fourth, the final *implementation* stage involves translating the identified pathways into tangible action plans. This stage necessitates defining intermediate milestones, allocating responsibilities and resources, and establishing mechanisms for progress monitoring.

Backcasting and truck-sharing

Because TS has not been successfully realised in practice or conceptually examined in research, backcasting is a suitable research method for bridging the gap between the status quo and the desired state. Compared with car-sharing concepts, which are increasingly adapted in practice, applying backcasting allows for a greenfield approach envisioning a potential conceptual implementation targeted at stakeholders such as freight forwarders and transportation companies.

The aim is to derive a development path towards the vision of TS by deriving specific milestones and recommendations for action. By identifying specific potential applications, different forms of TS can be identified and differentiated to develop a vision for the shared use of trucks according to use cases. From this vision of the future, we look back at the steps required to realise TS in the long term. The aim is not only to question practical reservations, but also to recognise and calculate the potential associated with TS.

According to the empirical research design, the approach translates to the identification of core milestones, such as business model design, and regulatory and platform requirements, to be bridged from a status quo perspective. Because of the novelty of the concept for practical usage, a linear forecast might yield differing conceptual outcomes, as illustrated in Fig. 3.2. Addressing uncertainty is a strength of the backcasting approach over other methods, because the process starts not with the status quo but with a future-oriented vision (Barrella & Amekudzi, 2011). Preliminary discussions with fleet operators have indicated uncertainty and scepticism towards sharing, because a TS concept seems far removed from the current typical truck deployment.

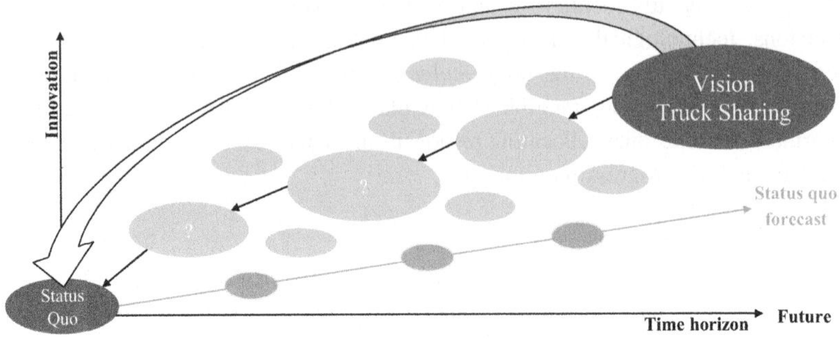

Fig. 3.2 Backcasting approach to truck-sharing

3.3 Research Design

Given the lack of explanatory findings from scientific research regarding why successful TS implementation is not yet observable in practice, this thesis initially focused on explorative research methods emphasising qualitative data collection. This approach is consistent with that of Miles and Huberman (1994), who have suggested that qualitative research is particularly suitable for early stages of investigation. Furthermore, to provide a practical contribution to the field, this thesis followed a pragmatic research approach to address meaningful real-world problems.

The empirical investigation of this thesis followed a mixed methods design involving various methods, and both qualitative and quantitative data collection and analysis approaches (Creswell & Clark, 2017). The empirical process is aligned with an exploratory sequential mixed method design, thus implying that the quantitative phase of data collection and analysis follows the qualitative phase.

Applied methods

In the research process, a set of methods was applied by combining semi-structured interviews, case studies, and an online survey as core method elements.

Element 1: semi-structured interviews

Given its novelty and limited observability in practice, investigating TS necessitates an approach able to accommodate the complexities inherent in this emerging field. Semi-structured interviews are particularly suitable for this type of research, because of their flexibility, and their ability to capture diverse and in-depth perspectives in the research phenomenon of TS. Thus, interviews were considered suitable for exploring new and complex topics, in agreement with King's (2004) description of interviews as "*the most common method of data gathering in qualitative research*" (p. 11). Interviews allow a wide range of perspectives from different stakeholders to be considered; this aspect is relevant because TS involves various contextual factors, such as economic considerations, regulatory environments, and social dynamics. For this emerging topic, interviewees might not have had pre-formed opinions or experiences. Semi-structured interviews thus facilitated the exploration of thoughts and attitudes towards TS. Integrating perspectives of LSPs, truck drivers, rental companies, associations, facilitators, and public authorities enabled a core set of questions to be combined and provided an opportunity to explore individual viewpoints.

Application: Interviews were conducted throughout the entire research process and across all three RQs. Initially, interviews were applied to identify possible application potential on the basis of the status-quo perspectives of practitioners followed by an explorative understanding of drivers and barriers regarding RQ2. Finally, interviews were conducted to validate the plausibility of the empirical results.

Element 2: case studies

Because this thesis describes the first investigation of concrete use cases for the application of TS in road freight transport, case studies are particularly relevant in bridging current truck deployment profiles to the vision of shared truck utilisation involving the perspective of practitioners. Case studies provide rich an in-depth data and information that reflect reality well and help capture the complexity of a TS concept. The case study approach followed the guidelines by Yin (2009) for how to perform case studies and involved quantitative data collection and analysis, as well as focus group interviews with truck fleet operators and additional interviews to deepen understanding.

Application: Case studies were used to investigate RQ1. Methodologically, the backcasting approach was not used for this RQ, in contrast to the other two RQs, because of the bottom-up approach aimed at identifying use cases by building on current weaknesses and inefficiencies.

Element 3: Online survey
A third element, an online survey, supplemented the previously conducted interviews and case studies. Transport companies, which ultimately decide on TS, and truck drivers, the stakeholder group whose work activities are most affected, were surveyed. This design enabled the inclusion of the perspectives of businesses and truck drivers, as key stakeholders, regarding the use of TS. Moreover, the two perspectives were able to be compared to reveal possible differences in perception. Specifically regarding truck drivers, the survey of deployment-related characteristics was intended to investigate whether typical patterns and driver-related distinguishing features might be relevant to the use cases. One advantage of the subsequent survey was that initial findings from the interviews were able to be deepened and enhanced with quantitative data in addition to qualitative information. Simultaneously, the use of open response fields enabled gathering of deeper insights for more differentiated analyses. In this way, derived hypotheses were tested and triangulated by combining them with interviews to verify consistency. The focus on driver- and truck-related characteristics placed the truck driver at the centre of the analysis to investigate the feasibility of TS and derive recommendations.

Application: The survey was aimed at generating insights into the drivers and barriers (RQ2) from various stakeholder perspectives. Alongside the case studies, this survey served as a foundational pillar for deriving actionable recommendations (RQ3).

Integrated research framework
The methodological considerations and elements introduced above were combined and translated into the research framework illustrated in Fig. 3.3, which focused on an exploratory sequential mixed methods design (Creswell & Clark, 2017) comprising semi-structured interviews, case studies, and a survey. In contrast to a pure sequential mixed methods design, the qualitative and the quantitative phase were not clearly separate, given the overlap between qualitative and quantitative elements in the survey. Specifically, the conducted survey also included qualitative elements underscoring the explorative focus of the thesis, and aimed at drawing on practitioner expertise to integrate perspectives. The survey design corresponded to a convergent mixed methods design enabling discussion of converging and diverging perceptions between entrepreneurial and truck driver perspectives.

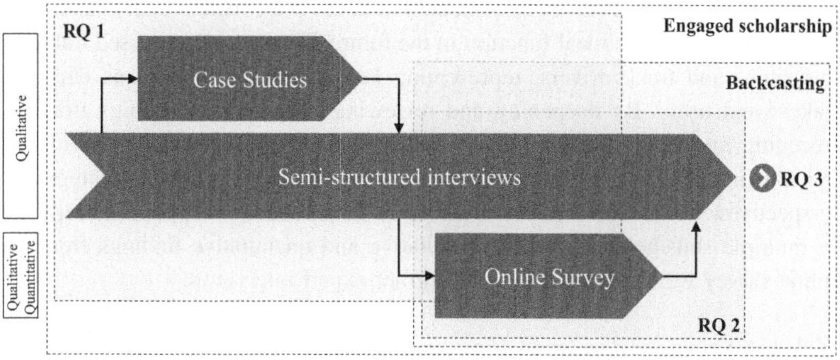

Fig. 3.3 Integrated research framework

As outlined in Sect. 3.2.1, the process adopts an overarching engaged scholarship approach to ensure practical relevance across all three RQs. Backcasting (see Sect. 3.2.2) is applied with regard to RQ 2 and RQ 3. Following a chronological perspective starting on the left side in Fig. 3.3, the research framework comprises three chronological steps in line with the three research questions posed in Sect. 2.6.

RQ1: Exploration of application potential

Investigation of RQ1 followed a two-step approach. The first step was examining potential applications of TS designed to create added value in practice. The industry-specific challenges from Sect. 1.1 formed the starting point. Interviews followed by case studies were initially used to identify current inefficiencies in the use and utilisation of trucks. To this end, efficiency measures (Sect. 2.1) were considered for truck utilisation. To ensure reference to transport planning, case studies and interviews in step 1 adopted an engaged scholarship perspective. Consequently, a concrete vision of TS was not predefined as intervention.

Second, initial preliminary findings were subsequently incorporated into the online survey in step 2. Hypotheses were tested quantitatively and supplemented with qualitative data.

RQ2: Exploration of drivers of, and barriers to, truck-sharing

On the basis of a backcasting approach, investigation of RQ2 focussed on the survey accompanied by pre- and post-interviews. During initial expert interviews and in the online survey, a visionary concept of TS was introduced, thus allowing interview partners and survey respondents to develop a concrete understanding of the TS concept and its ideal function in the future. The survey addressed transport companies and truck drivers, representing key stakeholder groups as decision-makers and users. By deepening and reviewing consolidated findings from the preceding interview stage and case studies, the survey comprised confirmatory elements in addition to exploratory elements. The design was aimed at integrating perspectives, because successful application in practice presupposes acceptance by multiple stakeholders. Finally, qualitative and quantitative findings from the online survey were validated through further expert interviews.

RQ3: Validation and derivation of recommendations
The investigation of RQ3 extended RQ1 and RQ2 through analysis from an overarching normative perspective focused on promoting TS. The derivation of recommendations targeted companies and public authorities able to exert influence.

3.4 Data Collection and Analysis

This section outlines the procedure of data collection and analysis across the three method elements—semi-structured interviews, case studies, and the online survey—conducted in this research.

3.4.1 Semi-Structured Interviews

Appropriate interview partners were selected through a purposeful sampling strategy, as suggested by Creswell (2007). The sampling process of interview partners followed a multi-step approach according to Saunders, Lewis, and Thornhill (2023). Acknowledging that TS involves and influences various stakeholder groups, as discussed in Chap. 2, a diverse range of interviewees from different stakeholder categories was intentionally selected. This approach was aimed at enhancing understanding of the research phenomenon and achieving triangulation. The interviewees were chosen according to a variation strategy, ensuring differences in organisational affiliation, organisational size, profession, age, and service scope.

An interview guide was constructed before the interviews to structure the interview process in a standardised manner. Adherence to the guide in every interview helped strengthen the reliability and minimise interviewer biases, because questions were asked in a more standardised manner than in open interviews (Gray, 2013). Because of the heterogeneity among interview partners, the guide comprehensively included seven sections with a broad range of questions. On the basis of information regarding the interviewees, selective adaption of interviews was followed to deepen topics while leaving out other sections, depending on the interviewees' knowledge.

Set time intervals between interviews allowed for further iterative development of the interview guide during the interview phase, according to the recommendations of Gioia, Corley, and Hamilton (2013). Newly raised topics were deepened, mirrored, and validated in subsequent interviews.

According to the backcasting approach, the following passage was presented to survey participants as an introductory framing intervention to TS:

> *In practice, it happens time and again that a truck is at a standstill or is only used for a (economically unprofitable) tour because it would otherwise be standing still. On the other hand, it would be practical to have an immediate replacement vehicle in the event of a good order situation or an unexpected breakdown. What contribution could truck-sharing make to this? Truck-sharing is essentially about sharing trucks across fleets with other companies. On the one hand, providers make vehicles available for a fee, customers select a suitable offer and use the truck for a rental fee. If a very large number of trucks are involved in sharing, the next sharing truck could always be nearby, no matter where you are. In the future, it is conceivable that a truck booked via a digital platform can be opened keylessly via an app and that the vehicle can also be returned after use without time-consuming administrative processes by uploading images of the interior and exterior of the vehicle as documentation in the app.*

In the early research stage, the depth of data rather than the number of interviews was considered essential. To achieve robust results, the interview process was continued until saturation of data was perceivable, according to a content-focused analysis. In total, 34 interviews were conducted, corresponding to the suggestion by Creswell (2007) of conducting 25–30 interviews in a heterogeneous population to reach saturation. Interviews lasted between 16 and 71 minutes and averaged 39 minutes. Interviews were performed via video conference calls or in person between May 2022 and June 2024. A list of interview partners and generic information is presented in Table 3.1.

Table 3.1 List of interview partners

No.	Organisation	Stakeholder group	Role	Country	Mode
1	Fleet operator/ FTL	LSP	CEO	GER	Workshop/ online
2	Fleet operator/ general cargo	LSP	CEO	CH	Workshop/ online
3	Fleet operator/ general cargo	LSP	CEO	CH	Workshop/ online
4	Fleet operator/ contract and retail logistics	LSP	Head of transport	CH	Workshop/ online
5	Fleet operator/ contract and retail logistics	LSP	Head of IT and digital transformation	CH	Workshop/ online
6	Asset-light freight forwarder	LSP	Sales manager	CH	Online
7	Job market platform/ logistics	Platform	Business development	GER	Online
8	Mobility software provider	IT	Senior account executive	GER	Online
9	Transport cooperation	Cooperation	Board member	GER	Online
10	Fleet operator/ general cargo	LSP	CEO	GER	Face-to-face
11	Fleet operator/ general cargo	LSP	Truck driver	GER	Face-to-face
12	Freight exchange platform	IT	Team lead partner and strategic account management	GER	Online
13	Fleet operator/ pharmaceutical logistics	LSP	Head of operations	GER	Online

(continued)

Table 3.1 (continued)

No.	Organisation	Stakeholder group	Role	Country	Mode
14	CEP service provider	LSP	Linehaul and network manager	CH	Online
15	Retailer/own account transport	Shipper	Head of transport logistics	CH	Online
16	Truck rental company	Rental	Branch operation manager	GER	Online
17	3PL/contract logistics, FTL	LSP	CEO	GER	Online
18	General cargo cooperation	Cooperation	CEO	CH	Face-to-face
19	Toll operator	Public company	Head of strategy and business development	GER	Online
20	Insurance company	Insurance	Head of market development	CH	Online
21	Fleet operator/ general cargo	LSP	Truck driver	GER	Online
22	Truck manufacturer	OEM	Sales manager/ key account management	CH	Face-to-face
23	Toll operator	Public company	Public sector and affairs/key accounts- and customer manager	GER	Online
24	Insurance company	Insurance	Head of motor fleet	CH	Online
25	Insurance company	Insurance	Head of motion	GER	Online
26	Fleet operator/ general cargo	LSP	CEO	CH	Online

(continued)

Table 3.1 (continued)

No.	Organisation	Stakeholder group	Role	Country	Mode
27	Express logistics provider/ healthcare logistics	LSP	Deputy CEO	GER	Online
28	3PL/3PL/ contract logistics, combined transport	LSP	Head of branch	GER	Online
29	Truck rental platform	Platform	Founder and CEO	GER	Online
30	Transport owner-operator	LSP/ truck driver	Founder and CEO	AT	Online
31	Logistics startup	Platform	CEO	GER	Online
32	Fleet operator/ FTL	LSP	Truck driver	GER	Online
33	Car sharing platform	Platform	CEO	CH	Online
34	Wholesaler/ own-account transport	Shipper	Fleet manager	GER	Online

Data analysis

Semi-structured interviews were recorded, and notes were taken during the interviews. For reasons of confidentiality, or to ensure an intimate atmosphere during the interview, several interviews could not be recorded. In those cases, only notes were taken. To ensure accuracy and correctness, and to validate the content and detect any misunderstandings, the recordings were reviewed after the interviews.

Data analysis was conducted according to Mayring's (2000) five-step method for content analysis. This approach is structured to systematically analyse interview data, ensuring that the findings are aligned with the research objectives. The steps involved in the content analysis were as follows: (1) developing a category system aligned with the research objectives, (2) coding relevant text passages according to this category system, (3) iteratively refining the classification framework, (4) re-coding the text by using the revised category system, and (5) interpreting and discussing the final results. In the exploratory sequential design setup of the interviews and online survey, qualitative content analysis was followed by validation interviews to ensure the robustness of the survey findings. These interviews served to triangulate interview and survey results. The data analysis software ATLAS.ti was used to support the content analysis.

3.4.2 Case Studies

As outlined in Sect. 2.1, the quantitative basis for truck utilisation is scarce. Therefore, data were collected empirically to generate insights into efficiency potential on the basis of actual truck utilisation of fleet operators, and to build a bridge to sharing. For evaluation of the utilisation of vehicles and efficiency potential, participating companies were asked to provide quantitative data on the key performance indicators (KPIs) shown in Table 3.2 for 1 year (2023), delivering data for 12 consecutive months. Development of the KPIs was performed iteratively together with three LSPs to ensure practical usefulness among truck fleet operators. Data collection was accompanied by interviews according to an interview guide and was supplemented by additional in-depth questions on truck utilisation, as summarised in Table 3.3.

Table 3.2 Focal quantitative KPIs of case studies on truck utilisation among fleet operators

KPI	Description
Empty trips	Share of empty trips as a percentage of a fleet's total truck trips. The length of a trip is not considered.
Empty kilometres	Share of distance travelled empty as a percentage of the total distance travelled by a truck fleet.
Truck utilisation	Load space utilisation rate in relation to all load kilometres travelled
Downtimes	Number of days on which trucks are not in operation despite being operational (excluding planned service days).
Operating time per day	Indication of the daily operating time of trucks, excluding downtimes in compliance with law conditions, e.g., Swiss night, public holiday, or weekend driving ban

Note: The KPIs are collected in relation to a company's entire fleet, not at the truck level.

Analysed KPIs

Empty running: An empty run occurs, for example, when a vehicle drives from the delivery of one order to the loading of the next without being loaded. In this type of operation, known as tramp transport, the proportion of empty runs can be as high as 50%. Empty runs result from asymmetric flow of goods and therefore are not always avoidable. Because the distance between delivery and collection can vary, empty runs should always be assessed in relation to the distance travelled. Depending on how the vehicles are used, the share of empty runs can be used to assess the efficiency of a logistics network in scheduled transport. The more often a vehicle can pick up a load simultaneously at an unloading point, the lower the proportion of empty runs. A high proportion of empty runs indicates negative efficient capacity utilisation but can be offset by a low proportion of empty kilometres, such as when loaded runs are made over long distances, but empty runs are made only over several kilometres. Because of the greater information value of empty kilometres, empty runs were not analysed.

Truck utilisation/load factor: Efficiency measures for truck fleet utilisation by McKinnon (2010) provide several approaches to measuring a truck's load factor, according to weight, volume, or a combined approach. In practice, no standard method is available to measure the load factor, and companies use their own definitions. Consequently, comparability among companies is limited. Nevertheless, the KPI is essential to capture the status quo of road freight utilisation, particularly given the scant public data available on truck utilisation in Europe (Sect. 2.1).

Downtimes: The term 'downtime' or idle days of trucks refers to the number of days on which trucks are available for use but are not in operation. Weekends and scheduled maintenance days are excluded. Inactive time can be considered idle capacity. TS, a direct mechanism for addressing this issue, has the potential to optimise fleet utilisation. Downtimes may occur for a variety of reasons, including truck driver absence, an insufficient number of orders, and administrative obstacles.

Table 3.3 Additional in-depth questions regarding truck utilisation in addition to quantitative data

Topic	Question
Quantitative data	• *Empty trips/kilometres*: What are the main causes of empty trips/kilometres? Are there company-specific reasons? • *Downtimes*: How do downtimes occur if trucks are ready for use? To what extent is downtime unavoidable? • *Operating time*: Could the utilisation intensity per vehicle be increased, at least for part of the fleet, such as through additional smaller tours, the use of a second crew, or lending to other companies?
Restriction capacity management	• What factors limit your transport capacity management and cause inefficiencies? • Which customer requirements particularly restrict your operational business? Where are you responsible?
Steering levers	• What steering parameters could you use to actively manage your capacities, particularly in times of increased demand volatility? • How do you balance over- and under-capacity? • What levers do you use to optimise capacity utilisation on individual routes? Do you see further room for improvement?
Optimisation potential	• What potential do you see for improving your own capacity management to increase the efficiency of tour planning? • How could potential be unlocked, and who/what is needed to do so?
Truck-sharing	• What are the indications of the daily operating times of trucks, excluding downtimes in compliance with law conditions (e.g., Swiss night, public holiday, or weekend driving ban)?

Note: The KPIs are collected in relation to a company's entire fleet, not at the truck level.

Companies involved in case studies
The case studies were initiated following a publicly funded consortial project[2] and were accompanied by workshops in which the results were discussed with practitioners involved in the project. Quantitative data were collected from five LSPs with differing business focus and company size. Company profiles are summarised in Table 3.4. Regarding company C, van operation for parcel delivery on the last mile was not considered, and the focus was on the companies' hub-to-hub-network deploying heavy-duty vehicles.

Table 3.4 Profiles of case study companies

Company	Business focus	Employees	Truck fleet size
A	Contract logistics, FTL/LTL	3,200	103
B	FTL/LTL, contract logistics	800	370
C	Parcel	1,200[3]	850[4]
D	General cargo, contract logistics	250	75
E	FTL	70	40

Fleet-related data were collected for each company and analysed across companies in a cross-case analysis. The quantitative results are presented in Sect. 4.1 in a diagram including all companies. Derivations and further comparisons were made on the basis of publicly available data and in-depth interviews.

3.4.3 Online Survey

Data for the online survey were collected through a structured questionnaire in German and English languages, which was designed to capture comprehensive information relevant to the RQs. The survey was implemented with the survey platform Unipark.

---

[2] INNOSUISSE project title "The concept of loadspace-shipment-sharing: at the example of a Swiss fleet operator community (LSS-FLOC) # perspectives towards truck sharing" (44422.1 IP-SBM).

[3] In Switzerland.

[4] Group-wide.

The survey was conducted by the Institute of Supply Chain Management, University of St.Gallen, in cooperation with the Federal Association of Road Haulage, Logistics and Disposal (BGL) e.V. and the job platform JOB-MATCH.ME. BGL shared the survey on the association's internal member platform, myBGL, to target the approximately 7,000 member companies. An additional 70 companies were invited directly through a personal approach and mailings. A call for participation was published in the online edition of the transport magazine *trans aktuell*. Truck drivers were recruited specifically via social media, through posts on Instagram, truck driver groups on Facebook, and LinkedIn. Flyers were also placed at a highly frequented truck wash on the German A5 motorway near Bad Rappenau (Cologne Truck Wash) to reach truck drivers waiting for their trucks to be washed. In addition, transport companies were asked to forward the questionnaire to their truck drivers.

For determining the required minimum sample size to achieve sufficient power to detect an effect for quantitative data analysis, a power analysis was conducted before the survey (Faul, Erdfelder, Lang & Buchner, 2007). G*Power 3.3 software was used to determine requirements for F tests to apply linear multiple regression analysis with three predictors. The following parameters were applied: 0.9 for the power test (Cohen, 1988), and 0.15 for a medium effect size (Faul et al., 2007). According to the calculation, the minimum sample was 108 cases. Data collection started on 21 March 2024, and lasted more than 2 months, ending on 31 May 2024. The timeframe aligned with recommendations by Creswell (2007), who has emphasised the need to ensure both comprehensive coverage of the research topic and a logistically feasible realisation. The subsequent data analysis was conducted in the statistics software SPSS version 29.

Data sample

The data cleansing was based on the criteria of completeness, uniformity, plausibility, duplicates, and missing values. Respondents who did not complete all relevant scales were excluded from the analyses. After the exclusion of 12 respondents, the total sample size was 113 participants, comprising 102 men and 11 women. The average participant age was 46.4 years, and the age ranged from 23 to 84 years (Table 3.5).

Table 3.5 Survey sample

	Option	Total number	Percentage
Gender	Male	102	90.5
	Female	11	9.5
Age	< 30	8	7.7
	30–39	27	23.9
	40–49	21	18.6
	50–59	25	22.1
	60 +	19	16.9
	Not given	13	11.5
Company turnover 2023	< 5 mio €	25	22.1
	6 to 25 mio €	21	18.6
	26 to 50 mio €	8	7.1
	51 to 100 mio €	5	4.4
	> 100 mio €	13	11.5
	Not given	41	36.3
Characteristics	Entrepreneur	50	44.2
	Self-employed	32	28.3
	Employee	61	54.0
	Active truck driver	61	54.0
	Freight dispatcher	17	15.0

Truck driver and deployment specifics
The 61 surveyed truck drivers had an average age of 43.91 years and therefore were slightly younger than the total sample; they had been working as drivers for an average of 18.2 years ($SD = 11.01$). Truck drivers' average covered transport distances were as follows: own-account transport, 3.29% ($SD = 12.00$); short-distance transport up to 50 km radius, 13.81% ($SD = 22.49$); regional transport between 51 and 150 km radius, 16.74% ($SD = 21.79$); national long-distance transport, 50.65% ($SD = 12.00$); and international long-distance transport, 14.52% ($SD = 23.22$). Because mean values do not indicate representative averages, dominant transport distances were calculated. Overall, approximately two-thirds of the surveyed truck drivers were active predominantly in long-distance transport (54.1% nationally and 11.5% internationally). Eleven truck drivers (18.0%) covered primarily regional transports, whereas only

ten truck drivers predominantly drove short-distance transport routes (13.1%) or own-account transports (3.3%). Accordingly, heavy-duty trucks with a payload of 18–40 t were deployed primarily for freight transport by 54 truck drivers (88.5%). Five truck drivers used primarily medium-duty trucks with a payload of 7.5–18 tonnes. Only two truck drivers operated with delivery vans with payloads as high as 3.5 tonnes.

Of the 61 truck drivers, 15 indicated never spending the night in the truck. Three truck drivers stated that they stayed overnight in the truck between one and five nights per month, ten truck drivers stated that they stayed overnight in the truck between three and five nights per month, and the remaining 32, just over half, reported staying overnight in the truck for more than five nights per month.

A total of 99 respondents provided information on company specific truck deployment and vehicle individualisation: 45 reported that the trucks were used exclusively by one driver, 24 reported that trucks were shared with exactly one colleague, and the remaining 30 respondents reported other usage profiles comprising alternating or flexible models.

Regarding the individualisation of vehicles to the needs of truck drivers, vehicles were fully individualised in 27 cases and highly individualised in 31 cases. A low level of vehicle individualisation was reported by 35 respondents, and no individualisation at all was reported by six respondents. Overall, the findings indicated a generally high consideration of truck drivers in the sample.

Empirical Research and Findings 4

Chapter 4 addresses the empirical research findings, according to the integrated research framework illustrated in Fig. 4.1. Section 4.1 delineates the potential applications and practice-driven use cases of TS, grounded in exploratory case studies (RQ1). Section 4.2 centres on identifying key drivers and barriers related to RQ2. The conclusion in Sect. 4.3 synthesises the empirical findings and derives implications and recommendations for action for stakeholders in the road freight transport sector.

4.1 Application Potential and use Cases for Truck-Sharing

In Sect. 2.1, the analysis of publicly available data on fleet utilisation indicated theoretically achievable potential for improvement in fleet utilisation, primarily through a reduction in empty kilometres and an increase in load space utilisation. However, publicly available data are limited and do not allow for more detailed differentiation among segments, transport types, vehicle sizes, and other parameters. Therefore, to draw conclusions regarding the application potential of TS, an in-depth case study analysis in Sect. 4.1.2 captures the status quo of individual fleet operators with different performance profiles and contributes to addressing

L. Häberle, *Truck-Sharing in Road Freight Transport*, Supply Chain Management,
https://doi.org/10.1007/978-3-658-48754-6_4

the challenges identified in Sect. 4.1.1. Finally, Sect. 4.1.3 translates application potential into feasible practical use cases.

4.1.1 Case Studies on Truck Fleet Utilisation: Status Quo Analysis

This section presents the findings of the five case studies conducted with transport companies. The quantitative data collected comprised the parameters specified in Table 3.2, which were demonstrated to be effective in measuring fleet utilisation performance. The objective of the quantitative analysis was twofold: to contribute to the limited data available on actual utilisation and to identify the extent to which efficiency potential could be exploited. This work provided a basis for presenting concrete use cases for the deployment of TS in Sect. 4.1.2, with an aim of addressing current inefficiencies.

Load space utilisation

The average utilisation of the load space over the year ranged from 80.5% for company D to 93.7% for company C. The values for all companies were significantly higher than those shown in Table 2.1, which indicates utilisation according to volume and weight. The discussion in Sect. 2.1 indicated that the values reported by McKinnon and Ge (2004) and Pahlén and Börjesson (2012), ranging from 52% to 64%, are theoretically correct but underestimate the actual utilisation of the load space. The reason for this underestimation is that a truck is considered fully loaded if either the permissible total weight is reached or the load space is utilised to capacity in terms of volume. Consequently, space may remain available, or the maximum weight might not be reached, yet a transport company records such a trip as fully utilised because loading additional freight is not possible.

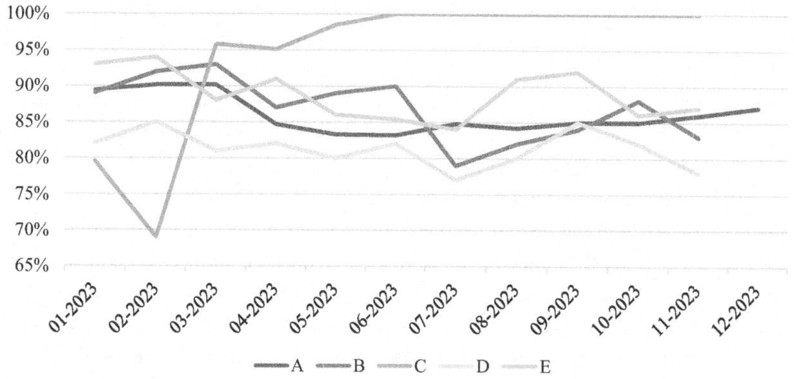

Fig. 4.1 Average truck utilisation across company fleets

A major challenge in the inter-company comparison of performance figures is the company-specific definition of tour utilisation. The disparate approaches to vehicle utilisation and the varying methods for measuring tour utilisation preclude a meaningful inter-company comparison. This aspect is illustrated by the utilisation figures of company C. The utilisation of swap bodies was contingent on the number of packages. The value, set at 100%, was determined by the company's controlling department, according to the expected composition of parcels from customers in the region. On the basis of years of experience, a specific value equated to 100% utilisation. Given the variability in parcel sizes, a vehicle considered to be 100% utilised according to the number of parcels might conceivably not have been at full capacity in practice. In contrast, the company also reported utilisation rates in excess of 100%. The values for 2023 indicated a marked increase in utilisation from June onwards, attributable to a logistics network adjustment. The closure of another location resulted in the inclusion of a greater number of smaller parcels in the main runs from June onwards. The influx of predominantly small parcels acquired during the summer from a Chinese online retailer led to a peak utilisation rate of 120% for the company. Although the utilisation rates were indeed high in the second half of the year, they did not reach the levels suggested by the figures.

Nonetheless, several insights can be derived. A focus on FTL transport is typically associated with high utilisation of loaded trips, and higher proportions of empty kilometres can occur as a side effect. In contrast, fixed-line transport, as exemplified by company D, exhibits lower utilisation rates, while the fixed timing results in fewer empty kilometres. Overall, tour utilisation must be contextualised according to the specific organisational environment and the company's resource configuration. The manner in which vehicles are utilised, in terms of distance range, number of shifts, type of cargo, tramp or scheduled traffic, and full or partial load logistics, significantly influences the utilisation of vehicles and the extent to which companies can influence this utilisation.

Interim conclusion: In practice, truck utilisation in road freight version is considerably better than suggested by the figures based purely on volume or weight. Nevertheless, the case studies with values between 80% and 93% indicated further potential for utilisation. Because a single transport company is embedded in logistics chains and usually cooperates with other companies, self-optimisation is not necessarily advantageous for optimising the overall transport network.

Empty running

In the case of empty kilometres, the focus is on the empty distances driven in relation to the total distance driven. The pure empty run share (in terms of counting the number of trips) was not evaluated, because these data would not provide meaningful insights. In the case of FTL transport, the number of empty runs might appear to be disproportionately high, even if only several kilometres are driven between unloading and the next loading. For instance, despite two full trips from A to B, and from C back to A, over several hundred kilometres, an empty run between B and C would result in a one-third proportion of empty runs. To avoid such distortions, only empty kilometres taking distance into account were recorded in the analysis.

As illustrated in Fig. 4.2, the empty kilometre share ranged from 8.5% to 25.0%. Two companies had the industry-wide average values of 23% and 26% in Germany and Switzerland, respectively. These figures were reported by the relevant authorities, KBA (2023) and BFS (2023a). The values for company B were notably low, fluctuating within a narrow range between 6.0% and 12.0% over the course of the year. The same rationale regarding capacity utilisation also applies in this case. Through a focused self-optimisation process, the company's own fleet was operated at optimal ratios, characterised by pairing of load flows. The primary activity was the delivery of goods to retailers, wherein empty containers, pallets, and other charge carriers could always be loaded as return loads. This modus operandi is distinct from that of companies specialising in FTL transport. In company A, the high service requirements resulting from just-in-time concepts for the reliable supply of production to automotive plants also played a role. Frequently, because of time restrictions, shipments on the same geographical route could not be consolidated.

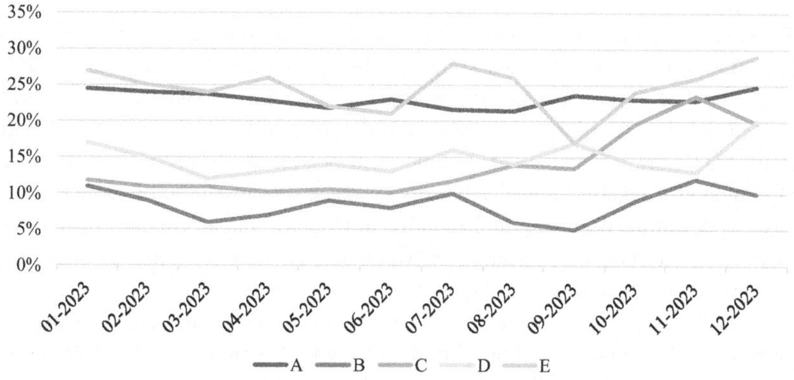

Fig. 4.2 Average empty running share across company fleets

Interim conclusion: In accordance with the findings regarding truck capacity utilisation, the participating companies also demonstrated efficiency superior to the average values reported for the overall transport market in terms of empty kilometres. Importantly, the companies included in the study were exclusively LSPs. The German Federal Association for Own Logistics & Shippers (BWVL) has stated that the own-account transport sector, which is represented by the association, accounts for 49.4% of all transport companies in Germany, equating to more than 46,000 companies, in addition to freight forwarders and pure transport companies. The provision of transport services is not typically the primary business focus of these companies. Assuming that optimised capacity utilisation is always the main priority would be erroneous. In contrast to freight forwarders and pure transport companies, in this case, other objectives have greater relevance, including the assurance of supply security, control over the logistics chain, quality characteristics, and even the visibility of vehicles as advertising media. An interview partner at a freight exchange substantiated this assertion in reference to search behaviour on freight exchange platforms. The example of a German kitchen manufacturer with its own truck fleet was provided, wherein the company avoids transporting backhaul loads to maintain a clean load zone; consequently, the notable increase in empty kilometres significantly influences the company's overall transportation costs.

Downtimes

The number of days per month during which the analysed fleets were not operational was relatively low, and showed a broad range of potential values. The mean number of days per month on which the system was unavailable slightly exceeded 1 day. As illustrated in Fig. 4.3, the mean values for companies A and B in 2023 were 0.5 and 0.7 downtime days per month, respectively. Company D exhibited a notably higher value of 2.9 days, and instances of more than 4 days of downtime occurring in 2 months. In the early months of 2022, company D made a substantial investment in expanding its fleet of vehicles, acquiring 20 new trucks in response to favourable market conditions and the emergence of capacity constraints in the freight sector. Because of unforeseen delays in delivery and a lengthy waiting period exceeding 1 year for new tractor units, the vehicles only gradually arrived in 2023, coinciding with a period of economic decline. The company therefore faced a challenge in maintaining high utilisation of its own fleet. Furthermore, the shortage of available truck drivers prevented the trucks from being utilised as frequently as usual. Companies A and B operated as freight forwarders, utilising only a portion of their own fleet, and therefore were able to excel in planning efficiency. A substantial proportion of the transportation was conducted by subcontractors. The high utilisation rates achieved by the company's own fleet resulted from a strategy prioritising the large order volume on its own trucks, to optimise capacity utilisation. In the event of a decline in transport volumes, the first step was reducing the number of subcontractors. Consequently, the two companies demonstrated minimal fluctuations throughout the year, and at times achieved consistent deployment on all working days within a given month. All five companies indicated that they operate in the capacity of both subcontractors and clients. However, the companies unanimously stated that a greater proportion of cargo was allocated to subcontractors than was acquired via the spot market and fixed contractual relationships with other freight forwarders. Furthermore, the case study partners highlighted that periods of inactivity did not inherently negatively influence the company's financial performance; this aspect was particularly true when opportunities were available to fulfil orders through subcontractors at lower cost, thereby leaving the company's own vehicles idle. However, from the perspective of the actors, a clear indication of unused capacity within the transport network was noted.

With an average of 21 working days per month, only 2 days without the use of a vehicle was equivalent to a vehicle downtime of almost 10% of all working days in a month. Given the exclusion of weekends, public holidays, and planned maintenance days from the analysis, potential for using trucks on more days was observed across the cases.

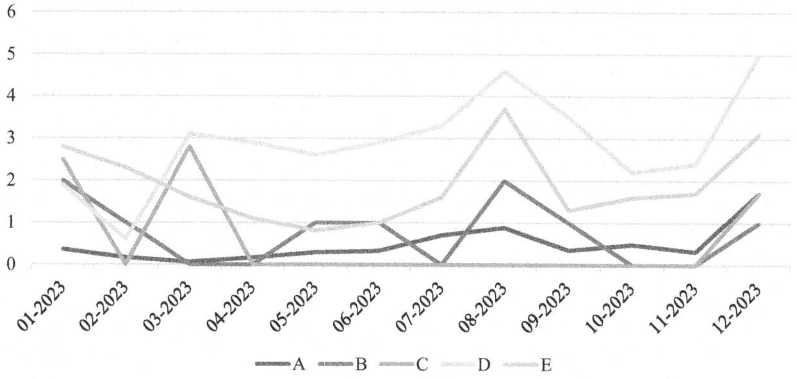

Fig. 4.3 Monthly truck downtime in days per truck

Regular servicing and repair work are imperative. Such maintenance may include routine procedures—such as oil and tyre changes, or brake inspections—as well as unforeseen repairs necessitated by mechanical faults. Additionally, regulatory considerations exist, such as the prohibition on operating trucks on public holidays and at weekends across Europe, or the more stringent restriction on driving at night in Switzerland. Additionally, the case study partners identified other potential causes, including a truck driver shortage due to factors such as holidays, illness, or vacancies, as well as a lack of orders or economically unprofitable orders. These phases may occur intermittently or persist over extended periods, according to the volatility of orders and the broader economic context.

The results of the in-depth interviews indicated that economic considerations influenced transportation companies' willingness to accept a vehicle in their yards for brief periods rather than having orders without available transportation. The following section examines the cost structure to contextualise this observation.

Downtimes and increasing depreciation costs for trucks in view of engine transition
The total costs incurred for the execution of truck transport operations were influenced by multiple cost drivers including both fixed and variable cost components. The relative importance of the individual cost drivers depended on the deployment profile and varied across countries and companies. For illustration, in the context of local transport, the utilisation of road infrastructure is associated with reduced toll costs, given the decreased mileage subject to tolls. In long-distance

transport, fuel costs are typically greater than those in shorter distances, whereas labour costs tend to be higher in labour-intensive shorter distances with numerous unloading points. Fig. 4.4 illustrates the average cost composition in German regional transport (up to approximately 150 km) in 2023. Wages constituted the largest cost block (34%) and were followed by fuel (20%) and truck costs (12%). The truck depreciation costs encompassed both fixed cost components for time-dependent depreciation and variable cost components for distance-related depreciation, including capital costs.

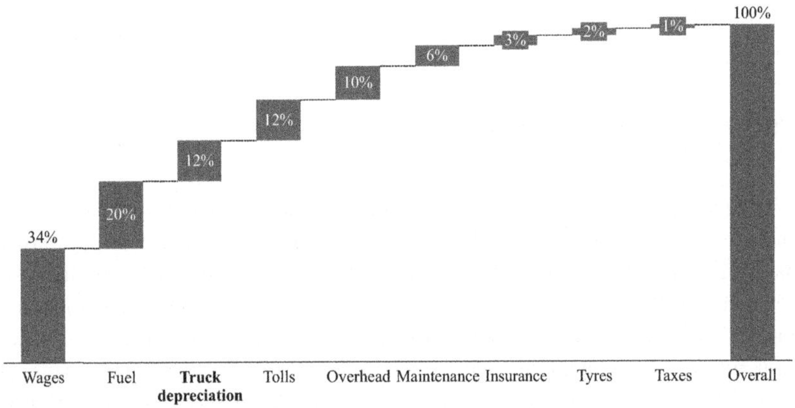

Fig. 4.4 Average cost composition for German national regional truck transport in 2023[1]

The proportion of costs for trucks varies over time, relative to the other cost drivers. Table 4.1 summarises the results from various sources, which have estimated depreciation costs between 8.8% and 14.8%; 12% was considered a plausible indicative value.

The costs associated with the trucks, such as wages, are part of the costs of readiness to perform, which are incurred largely independently of vehicle utilisation. Although fuel costs are higher than truck depreciation costs, they arise only when the vehicle is in use. The derivative nature of the demand for freight transport results in limited flexibility among LSPs to align the size of their fleet with transport demand in the short term. Freight exchanges are therefore widely used for balancing loads. When a transport company has the option of accessing capacity as required through a sharing offer, at least some of the vehicle-related fixed

[1] Based on BGL (2024) and conducted case studies.

costs can be made variable, thereby increasing flexibility, including in financial terms.

Table 4.1 Relative truck depreciation costs in total road freight operating costs

Year	Country	Depreciation	Distance profile	Reference
1997	Austria	14,8%	Long-distance	Arbeitskammer Wien (2001)
2000	Austria	9,0%	Short-distance	Arbeitskammer Wien (2001)
2000	EU-15	10,9%		Musso (2001)
2016	Czech Republic	13,0%		Kampf, Potkány, Krajčírová & Marcineková (2016)
2023	Germany	8,84%	Long distance	BGL (2024)
2023	Germany	11,73%	Regional	BGL (2024)
2023	Germany	13,04%	Short-distance	BGL (2024)

Shifting cost structure

As a consequence of changes in drive systems, a shift in the cost profile illustrated in Fig. 4.4 is anticipated in future years. Projections indicate that the future market share of alternative drive systems will increase. According to their technology roadmaps, the leading original equipment manufacturers (OEMs), namely Scania, Renault, MAN, DAF, Volvo, Mercedes-Benz, and Iveco, plan to increase the proportion of new entry emission-free heavy commercial vehicles to between 20% and 60% by 2030.

Transport companies anticipate a transformation in the cost structure of zero-emission vehicles, whereby the acquisition costs will be significantly lower than the status quo in the long term but will remain higher than those of a comparable conventional diesel engine. At present, transport companies are offered market prices for emission-free vehicles that are two to three times higher than those for a comparable conventional diesel engine. In the third quarter of 2024, transport companies reported that the price of an electric tractor in Germany is slowly decreasing to €200,000, more than twice that of a diesel tractor unit (Landwehr & Loew, 2024). Nevertheless, the higher acquisition costs are offset by reduced consumption costs, which are amortised over time.

From an economic standpoint, in considering the total cost of ownership over the vehicle's lifespan, emission-free trucks become increasingly attractive in scenarios in which the annual mileage is high. As a consequence of the increased initial investment required for the vehicle, the opportunity costs associated with non-moving trucks increase in line with the drive change; consequently, greater significance is attributed to non-moving trucks. Interviews with industry experts indicated that avoiding downtime will become a more significant factor in the deployment of electric trucks. The current profiles of electric truck deployment reflect this trend. Companies have stated that they already utilise electric trucks, thus achieving high mileage and minimising downtime. Consequently, the transition towards emission-free trucks acts as a driver of change making sharing more attractive to reduce downtime.

Conclusion: Case studies indicated that the analysed companies achieved better capacity utilisation across their truck fleet than that typically reported in the literature for the industry. In-depth interviews suggested that this finding might also apply to the industry as a whole. Nonetheless, the results indicated that improvements are possible in all efficiency dimensions analysed, including less downtime, fewer empty kilometres, and higher load capacity utilisation. The risk of increased downtime and load space utilisation fluctuations was particularly high for transport companies that work as subcontractors for (usually larger) freight forwarders. Interviews also suggested that the efficiency potential in own-account transport is likely to be higher compared to LSPs, because the provision of transport services is not a core competency of these companies.

4.1.2 From Truck Utilisation Inefficiencies to Application Potential

This section discusses factors limiting higher transport efficiency of trucks in road freight transport, according to the experience of the five case study companies, and outlines improvement potential that could be addressed by the adoption of a TS concept. In-depth interviews were conducted to identify the reasons preventing better fleet utilisation. The factors derived from the literature in Table 2.2 were confirmed, including unpaired freight flows, demand structure from the customer side, operational constraints, route planning, economic pressure, customer requirements, and freight specifics.

All companies inevitably experience unpredictable volatility in their transport volumes. Notably, only a small proportion of customers exhibited consistent transport demand, thus resulting in fluctuations throughout the year. One reason for this finding is that the provision of transport services gives rise to a phenomenon referred to as derivative demand, which results from the physical movement of goods within the economy. The provision of freight transport services is a direct consequence of economic activity. Thus, the deployment of personnel and vehicles is not scheduled according to a fixed plan but depends on the demand situation. Because freight transport services are intangible and are not subject to storage, the decision-making process for freight transport service providers involves determining the optimal allocation of resources to meet the identified demand. A focus on peak demand may result in costs associated with idle capacity during periods of reduced demand. In practice, the phenomenon described in Sect. 4.1.1 is evident: transport companies are more willing to accept a certain degree of costs for downtime and empty kilometres to provide customers with reliable transport services. The choice of larger capacities entails higher fixed costs. The acquisition of supplementary transport capacities from third parties, for instance at short notice, through rental or situationally involving subcontractors via the spot market, may result in variability in these costs in comparison to the maintenance of a company's own capacities.

What improvements did the case study partners perceive? Company A considered itself to be operating close to the optimum level for all KPIs, and therefore did not perceive significant potential for further improvement, primarily because of the strong focus on self-optimisation, accompanied by the outsourcing of all tours and shipments that could not be optimally served by the company's own fleet. The routes can be assumed to be less well utilised by subcontractors. The situation differed for companies B, C, D, and E, which had higher proportions of empty kilometres in their fleets. With the exception of company C, the utilisation rates were lower, and all companies had more days of inactivity. In light of these observations, the online survey was aimed at investigating the potential contribution of TS.

Survey-based truck-sharing application potential

According to the literature on asset sharing, flexibility is considered a major motive for sharing initiatives and thus was a core element of the TS conceptual fundamentals outlined in Sect. 2.4.2. The importance of flexibility as an opportunity for sharing was emphasised qualitatively in the interviews. In addition, further dimensions were collected to enable quantitative investigation. Nine items were developed, which were surveyed on a 7-point Likert scale. All items were generated inductively from the interviews and focused on relevant pressure points in road freight transport to be addressed by TS. To ensure that no relevant dimension was excluded, participants were able to provide additional information in an open response field. No further dimensions were derived from this process, thus indicating that the nine items covered industry pressure points comprehensively.

The results are summarised in Table 4.2. The greatest potential was associated with reducing downtime ($M = 4.79$), followed by increasing flexibility ($M = 4.77$). No significant differences were found between groups of entrepreneurs and non-entrepreneurs, or between truck drivers and non-drivers.

Table 4.2 Variable profile: rating scale truck-sharing potential

Truck-sharing can contribute to…	*M*	*SD*
1) optimising **load space utilisation** of trucks.	4.09	1.87
2) increasing the **flexibility** of transport companies when deploying the fleet.	4.77	1.75
3) reducing **empty runs.**	4.05	1.82
4) reducing **transport costs.**	4.11	1.87
5) reducing **carbon emissions** from transport.	3.69	1.89
6) increasing the **quality** of transport services provided.	3.16	1.75
7) increasing the **reliability** of supply chains.	3.71	1.80
8) reducing the **downtime** of trucks.	4.79	1.91
9) improving the **profitability** of transport companies.	4.32	1.80

Note: Items were rated on a 7-point Likert scale.

The other two efficiency factors, utilisation and empty runs, scored lower than downtime, with mean values of 4.09 and 4.05, respectively, ranking in the middle of the scale. Although the exchange of vehicles can directly lead to the avoidance of downtime, the respondents considered the potential for sharing to also lead to higher utilisation and fewer empty kilometres to be less apparent. Good route planning is crucial. Thus, the potential for reducing carbon emissions was rated fairly low ($M = 3.69$).

Item 6 had the lowest mean value, at 3.16. Accordingly, quality was considered a critical point to which little potential was attributed. Interview partners expressed scepticism regarding quality. Although some interview partners took a neutral stance, none of the experts saw any advantages of sharing in terms of transport quality. This finding was also reflected by the low score of 3.71 for item 7. The lack of care in handling the property of others was considered a latent risk in sharing. Insurers confirmed that the loss ratios for shared vehicles were significantly higher than those for privately owned vehicles.

Increased flexibility and reduced downtimes

In light of the derivative nature of demand for freight transport, and the concurrent production and consumption of transport services, reactivity has a critical role. The capacity utilisation of companies is optimised through planning in an expedient and flexible manner. In this regard, the potential for flexibility and downtime are linked. The practice of TS has the potential to enhance the flexibility of truck utilisation, by allowing companies to rent vehicles on an as-needed basis and at short notice, rather than maintaining their own fleet of trucks. In contrast, vehicles can be made available on the market during periods of low demand. This aspect is particularly advantageous in the case of fluctuating transport volumes or seasonal peaks, by avoiding overcapacity. TS therefore helps reduce downtime, because several users can utilise the trucks more efficiently, thus leading to better utilisation of the vehicle fleet and minimising unused time. Although TS does not directly result in increased capacity utilisation or a reduction in empty kilometres, the associated increase in flexibility indirectly promotes these efficiency measures. The full chain of effects is schematically illustrated in Fig. 4.5.

Estimating the overall vehicle reduction potential

To address the societal perspective in addition to the company-centred perspective, the interview partners were asked to estimate the potential for reducing the number of trucks on the roads that they considered possible if TS were to be established. A total of 17 participants provided a quantitative estimate. The mean estimate was 2.6% (SD = 2.87), and the range was between 0% and 12%. When asked about the differences with respect to cars, all interview partners stated that they considered the potential to be lower for trucks. The main reason given was that cars are on average much less utilised than trucks, whereas the optimisation requirement in commercially operated freight transport is high, and the efficiency figures are correspondingly much higher. This assessment is consistent with studies on the average seat utilisation of cars. According to Herrmann, Jungwirth, and Huber (2022), cars have an average seating capacity utilisation of 1.1%, on the basis of a theoretical maximum of five seats per car being used for 24 hours.

Although not directly comparable to the results of the case studies, this low utilisation rate provides a suitable explanation as to why the idea of sharing cars evolved earlier than that of sharing heavy trucks in road freight transport.

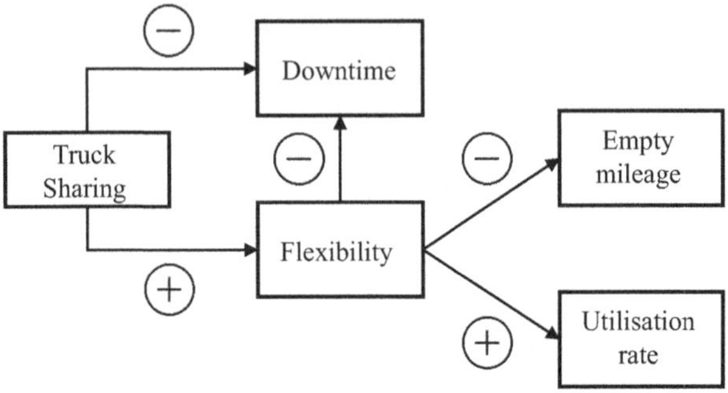

Fig. 4.5 Effect chain of truck-sharing

4.1.3 Use Cases for Truck-Sharing in Road Freight Transport

This section presents the research findings focused on use cases for TS in road freight transport. The outlined use cases are derived from in-depth interviews with industry experts and key stakeholders, and build on case study insights presented in Sect. 4.1.1. Use cases address various circumstances in which TS could be effectively integrated into and add value to current transport operations. The investigation followed an inductive approach based on specifically described use cases from stakeholders that were validated with further interviewees.

Table 4.3 lists use cases that were generated bottom-up and then summarised into higher-level categories. In addition to the described example, passages extracted from the interviews are documented. The use cases were identified and confirmed to be typical of the challenges that TS might address and fit the category.

Table 4.3 Overview of identified use cases for truck-sharing derived from expert interviews

Category	Example	Extracted interview quotation (translated)
Seasonality	Strawberry and asparagus harvest	*"Our season lasts only about 12 weeks, depending on the weather, and that's exactly when we need storage and transport capacity."*
	Grape harvest	*"In our Wachau region, the grape harvest in autumn is a peak period. We have been doing this with a winemaker friend for years, lending him a truck for 2 to 3 weeks."*
	Distribution logistics for do-it-yourself stores	*"Distribution logistics for do-it-yourself stores is typically highly cyclical. We see the biggest peak in spring, around March transporting more than double the shipments compared to the weakest months. It is particularly quiet in summer and at the beginning of the year right after Christmas. That's when we have this dedicated truck capacity available."*
Volatility	Balancing fluctuating supply and demand	*"If we reliably had the option of spontaneously renting a truck for a short period of a few days or lending it out when the order volume is low, we could better deal with fluctuations."*

(continued)

Table 4.3 (continued)

Category	Example	Extracted interview quotation (translated)
Flexibility	Variable, demand-driven fleet size	*"Especially among us SMEs, you quickly notice when a major customer drops out or joins. We then adjust the size of our fleet upwards or downwards accordingly. We have experience with long-term rentals of 1 or 2 years in order to be more flexible compared to the strategy of relying solely on purchased and leased vehicles. However, for greater flexibility, you also have to accept slightly higher costs for the trucks. Sharing could therefore be a useful addition."*
	Market responsiveness	*"Ultimately, as a transport company, we are heavily dependent on the market situation. With our fixed fleet, we react to this as best we can and have to use the spot market to find loads, which sometimes means low pay. Companies that don't have a fleet and get capacity as needed are at an advantage. […] The big freight forwarders all do it this way, they hardly run their own fleet and outsource the risk to their smaller subcontractors."*
Costs	Fixed cost reduction	*"Assuming that we could reduce our fleet of currently 70 trucks by just 10 units by sharing them on the sharing market, we could already reduce our high fixed costs in the fleet considerably. That would be a noticeable financial relief."*

(continued)

Table 4.3 (continued)

Category	Example	Extracted interview quotation (translated)
	New profit centre	*"Costs are one aspect, revenues are another. Having the option to situationally lend out a truck that is not needed at the moment, opens up a new source of revenue for us as a fleet operator."*
Vehicle specificity	Access to specialised truck equipment	*"The more specialised the vehicle, the more likely it is to remain unused in the yard. Apart from standard semi-trailers, there are also special vehicles that are only used in niche areas, such as heavy transporters. We would never buy a liquid tanker either, as we would not be able to utilise these vehicles to full capacity on a permanent basis with our customer structure. A large pool of suppliers and customers would have the advantage of being able to access special vehicles when needed."*
Vehicles with alternative drive systems	Facilitating mobility transition Reducing downtimes	*"The business case for alternative drive systems is different: they are costly to purchase, but cheaper to operate in the long term. That's why we use our electric trucks on the routes with the longest distances and have practically no downtime. We pay much more attention to downtimes than we do with diesel trucks. We reach the break-even point after seven years."*

(continued)

Table 4.3 (continued)

Category	Example	Extracted interview quotation (translated)
	Testing	*"Large fleet operators are better placed to integrate a few alternative vehicles into their fleets. However, those who operate only a few or even a single truck will think twice about whether they can afford to switch at all considering that the vehicle is two to three times the price. I therefore see testing and gaining experience with hydrogen or electric trucks as an opportunity for small companies in particular to achieve a broad-based shift in drive technology."*
Deployment intensity	Multi-shift operation to reduce downtime	*"Many transport operations are well plannable and regular, in regional transport sometimes only running in the morning for 3 hours. Others are on the road in the afternoon. Bilaterally, it would be much too inefficient to find a compatible partner to use a truck with another company. Via a platform with a matching algorithm, it looks quite different. Perhaps not so much for us freight forwarders as transport specialists, but especially for own-account transport operations there is potential for permanent sharing (fixed days of the week or changing times of day). I mean companies that do not operate transport as their core business, they often have more downtimes."*

Note: Quotations were selected to describe a typical use case for the corresponding category.

Among all collected use cases, seven categories were identified: seasonality, volatility, flexibility, costs, vehicle specificity, vehicles with alternative drive systems, and deployment intensity.

Seasonality: Because of the derivative nature of demand for freight transport, pure transport companies are dependent on the demand situation on the customer side. For pure transport companies, fluctuations are a challenge that limits alignment of fleet size with transport demand in the short term. In practice, the demand for goods transport is usually subject to seasonal fluctuations. Specific sectors such as agriculture are particularly affected. For example, harvest cycles result in a punctual demand in only several weeks of the year (Table 4.3). The demand behaviour of customers in the construction industry also fluctuates greatly over the course of the year. One interviewee described the distribution cycles for a customer in Germany, wherein more than 3,000 pallets per month must be transported during the peak season in spring, whereas in the slowest months, fewer than 1,400 pallets, less than half that amount, must be transported. Because the goods are seasonal, shifting the transport to earlier or later times is scarcely possible.

The amplitude of fluctuations is a problem for transport companies with an average high utilisation rate if fluctuations in the demand profile cannot be offset by a heterogeneous customer structure. Many LSPs therefore have a broad strategic base and serve different industries. However, many companies deliberately operate in niche markets to compete with a small number of competitors and consequently can position themselves as experts and better serve specific customer needs.

The sharing of trucks offers an opportunity to add assets in high-volume months to cope with pronounced fluctuations, whereas assets can be offered on the market outside of seasonal periods with low demand. This principle functions well when the periods among industries complement one another. However, greater customer and industry diversity are associated with greater likelihood that supply and demand can be balanced on the market.

Volatility: In addition to seasonal fluctuations, which can largely be anticipated, unsystematic fluctuations also occur. Such fluctuations may be attributed to economic cycles and are often associated with a general increase or decrease in transport volumes. Economic growth is typically associated with an increase in transport demand, whereas periods of recession are usually characterised by a decline in demand. This aspect is demonstrated by the German truck toll mileage index, which serves as an early indicator of economic development (Destatis, 2024). The sharing principle necessitates the convergence of supply and demand. In this regard, the TS concept is particularly well-suited to situations in which volatility occurs in an unsystematic manner, affecting only a portion of the market at a given time. Large economic cycles can be assumed to lead to stronger representation of either the supply or demand side. However, importantly, volatility also occurs in individual sectors. For example, the global pandemic caused by the SARS-CoV-2 virus led to a sharp increase in transport volumes in healthcare logistics as a result of the implementation of testing capacities and vaccination programmes. In contrast, other sectors did not experience a similar boom during this period.

Flexibility: The practice of TS can assist transport companies in achieving greater flexibility in fleet management by enabling the balancing of over- or under-capacity, resulting from seasonal or volatile factors. Both OEMs and transport companies have independently confirmed that purchasing and leasing continue to account for the largest proportion of vehicle acquisitions, whereas renting remains a significantly less common practice. One principal reason is the higher costs associated with renting rather than purchasing or leasing vehicles. In contrast, optimising the utilisation of a fixed fleet is challenging and could pose risks for transport companies if they were to experience a substantial reduction in their transportation volumes, for instance as a result of the termination of a customer relationship. Fleet operators might plausibly use TS as a targeted control element with the objective of ensuring that capacity is flexible.

This can directly result in a reduction in downtime. Furthermore, the interviews indicated potential for indirectly increasing utilisation and reducing empty kilometres if more optimal use cases for vehicles could be identified. An illustrative example is the functional principle of apartment swapping via platforms such as Airbnb. When individuals embark on a period of leisure, they might seek accommodations in another region and concurrently rent out their own property, to address the need for accommodation, for instance, during a business appointment; consequently, the two parties exchange the apartment, and both parties benefit. This concept can be extended beyond a dyadic relationship to include many parties.

Costs: Because freight transport services cannot be produced in advance, the expected transport capacities must be available in case they are needed. The provision of trucks involves fixed costs. In light of the low margins in the industry, combined with high liquidity requirements, the option of reducing part of the fleet and relying on shared vehicles instead promises substantial financial relief. Beyond cost reduction, new revenue streams might be exploited if unused assets are leased out.

Vehicle specificity: A further potential application of TS in road freight transport is the provision of access to specialised equipment for fleet operators. In many instances, particularly among smaller fleet operators, the financial resources or consistent demand may be insufficient to justify the ownership or long-term leasing of trucks equipped with specialised features, such as refrigerated units, flatbed trucks, or tankers. TS may provide a cost-effective and flexible solution to this challenge, by enabling users to fulfil specific customer requirements or accommodate unique shipments without the need to maintain their own fleet of specialised vehicles.

For example, a fleet operator whose primary business is the transport of general cargo may occasionally require the use of a refrigerated vehicle to facilitate the movement of perishable goods. Rather than investing in a refrigerated truck that would remain underutilised, the operator could utilise a TS service to rent the vehicle only when necessary. This use case would also benefit the owners of specialised equipment, who could increase the utilisation rates of their assets by making them available to other operators. By sharing their specialised trucks, these owners could generate additional revenue, thereby offsetting the costs associated with owning and maintaining high-value equipment.

Vehicles with alternative drive systems: TS has been identified as a potential enabler for the adoption of alternative drive system vehicles, including electric and hydrogen-powered trucks. This aspect is particularly relevant in the context of smaller transport companies. Such firms frequently lack the financial resources to invest in advanced technologies, because of the considerable upfront costs involved (Landwehr & Loew, 2024). The deployment of a single vehicle, or a small number of vehicles, or the acquisition of a vehicle costing twice the price or more than a comparable diesel truck, might exceed the financial capabilities of such firms.

The utilisation of TS platforms enables fleet operators to access these vehicles without a need for significant investment, thus facilitating market penetration. For example, a company that utilises predominantly diesel trucks might rent an electric vehicle for urban deliveries, thereby ensuring compliance with low-emission zones. Moreover, this approach provides practical experience with alternative drive systems, thus enabling operators to evaluate factors such as charging infrastructure and maintenance requirements before committing to long-term investments.

Deployment intensity: TS has the potential to increase vehicle deployment intensity in road freight transport, by enabling practices such as double shifts, thereby reducing downtime between operations. This approach enables the efficient pairing of companies with complementary schedules and consequently maximises vehicle utilisation. For example, a truck might be utilised by one company in the morning and another in the afternoon, or alternatively might be utilised for part of the day and then experience downtime during other hours or days. One example might be a logistics provider that shares its fleet with a partner company. The partner utilises the trucks during off-peak hours, thus effectively doubling the vehicle's daily operational time and minimising periods of inactivity. Further case studies have indicated that such utilisation patterns are common. In these cases, TS optimises operational time, minimises downtime periods, and significantly enhances overall efficiency. Sharing limited to several hours is the optimal solution, provided that regular patterns are identified, thereby enabling geographically close sharing partners to establish regular sharing.

The realisation of the identified potential and concrete use cases in practice is determined by a multitude of influencing factors. In addition to functional requirements, user acceptance is a crucial factor in the transition of potential into practice. Therefore, drivers and barriers were analysed in this thesis, to provide insights for designing the business model and operator model. The results are presented in the following Sect. 4.2.

4.2 Drivers and Barriers to Truck-Sharing

This section identifies the barriers to TS implementation. First, the barriers are categorised and classified according to their applicability to the relevant parties, to provide a foundation for transforming barriers into drivers. Section 4.2.2 provides a detailed analysis of the implications of truck driver sharing in labour law. Section 4.2.3 examines the IU to assess the current level of acceptance of TS in practice. The section concludes with the derivation of deployment profiles with increased openness to TS in Sect. 4.2.4, which identifies acceptance-related patterns. This work provides the basis for the derivation of implications and recommendations in Sect. 4.3.

4.2.1 Identified Barriers to Sharing Trucks in Road Freight Transport

Barriers were identified according to the findings of the qualitative interviews, which were aimed at elucidating the nature of barriers at multiple levels, from the perspectives of various stakeholders. The identified barriers can be classified into two categories. First, *functional barriers* are understood as a prerequisite for sharing. These barriers, including compliance with legal framework conditions, must be overcome for a sharing transaction to be feasible. Additionally, functional barriers are relevant from an efficiency perspective and must be addressed for sharing to generate added value beyond that of the status quo. Second, *perceptual barriers*, independently of the other two categories, reflect attitudes towards TS at the individual level. Table 4.4 describes the findings of the analysis of functional barriers, whereas Table 4.5 illustrates the perception-related barriers.

Functional barriers

At the functional level, several factors are linked, including the alignment of supply and demand, the positioning and suitability of vehicles, and an accepted refuelling process. A transaction can proceed only after these fundamental prerequisites have been satisfied. For example, companies must have a reliable vehicle availability plan in place. They seek to avoid becoming overly dependent on a concept that might not perform as expected. The anticipated cost savings, such as reduced fixed costs, can be realised only if the fleet is optimally reduced. Furthermore, a TS concept must comply with the applicable legal framework conditions. In addition to the obligation to provide proof of vehicle insurance, clarifying national toll legislation and, if truck drivers are included in the sharing process, labour law issues, is essential. After the procedural requirements and legal compliance have been ensured, the business framework represents a further functional barrier. From a cost perspective, the necessary handover process for each sharing transaction is an inefficient use of resources. Experience from the traditional truck rental business indicated that the duration of the rental limits the financial attractiveness if manual handover and instruction processes are required. In addition, companies that rent out vehicles are concerned about bearing the costs of damage to vehicles.

Perceptual barriers

Besides functional barriers, perceptual barriers have emerged. Change management projects demonstrate that people's opinions and perceptions are pivotal in driving change. In practice, success in change management projects is not solely determined by functional rationale. From an entrepreneurial perspective, the lack of acceptance of TS by employed truck drivers is a business risk of losing truck drivers. It is also indispensable for companies to recognise the value that TS creates. The perception-related barriers are recorded and categorised against this background.

Table 4.4 Functional barriers to truck-sharing

Sub-category	Description
Supply-demand match	Supply and demand must match from a temporal perspective for a sharing transaction to occur. The initial ramp-up phase involving many vehicles poses a particular risk.
Vehicle positioning	Geographical alignment of supply/demand is necessary to utilise vehicles with minimal empty kilometres. A suitable location for taking over and parking vehicles must be provided for both parties.
Vehicle feasibility	The functional requirements of trucks are more heterogeneous than those of cars. Depending on the type of freight, specific vehicles may be required.
Refuelling	Ensuring cost allocation for refuelling/charging, on the basis of the principle of causation, and ensuring that a minimum fuel or battery charge level is maintained when the vehicle is returned are necessary to avoid delays and downtimes.
Labour law	If truck drivers are also part of the sharing concept, beyond vehicles, additional labour law frameworks must be complied with (discussed in Sect. 4.2.2)
Vehicle insurance	By default, fleet operators' vehicle insurance usually does not cover rental cases. Insurers confirm increased loss ratios for sharing, which, in addition to a lack of risk assessment, acts as a cost driver in sharing.
Toll accounting and liability	Because vehicle owners are liable for vehicle-related toll payments, they face a potential financial risk of liability for non-payment by users renting a vehicle.
Claims allocation	The cross-company deployment of vehicles by multiple truck drivers poses challenges in assigning damages and other claims to the appropriate party.
Vehicle handover process	The shorter the sharing period, the greater the cost of manual processes. In the established truck rental business, the complexity of the manual handover processes is a substantial financial obstacle for rental companies offering short-term rentals.

Table 4.5 Perceptual barriers to truck-sharing

Sub-category	Description
Vehicle cleanliness	Concerns exist regarding users not keeping vehicles clean in accordance with their own standards, both inside and outside the cabin.
Commuting time	The practice of sharing should not increase the time required for truck drivers to reach their trucks. The ability to plan working hours is a related pressure point for employed truck drivers.
Storing personal belongings	The practice of frequently changing between vehicles can be a source of discomfort for truck drivers that necessitates the clearance of personal items. This practice can give rise to a perception among truck drivers that their private sphere is being encroached upon if other people use ‚their' trucks.
Firm branding on vehicles	A company's truck is often perceived as advertising for the company itself. Alternatively, companies often contract with third parties to drive their own branded vehicles. In the case of sharing, however, the owner does not influence the driver's behaviour. Using a vehicle bearing a direct competitor's logo is perceived as particularly critical.
Familiarity with vehicle	Truck drivers might face challenges in adapting to new vehicles, thus potentially posing safety risks, decreasing comfort and efficiency, and leading to confusion and potential errors. Each truck has unique driving characteristics, control systems, and assistance technologies that can complicate the transition between vehicles.

Classification of the derived barriers at the company and truck driver level (Table 4.6) indicated several patterns. On the one hand, the functional barriers were all relevant at the company level, and only several of them affected truck drivers directly. On the other hand, the situation differs because of perceptual barriers primarily concerning truck drivers. Although the well-being and satisfaction of truck drivers must be assumed to be highly important to companies, particularly given the truck driver shortage (Kille et al., 2023), the five perceptual barriers in Table 4.6 relate predominantly to truck drivers at the individual level. Thus, the various levels of the system must be addressed to overcome the identified barriers.

Table 4.6 Classification of barriers at the company and driver levels

Barrier	Company	Truck driver
F: Supply-demand match	**X**	
F: Vehicle positioning	**X**	**X**
F: Vehicle feasibility	**X**	
F: Refuelling	**X**	
F: Labour law	**X**	
F: Vehicle insurance	**X**	
F: Toll accounting and liability	**X**	
F: Claims allocation	**X**	
F: Vehicle handover process	**X**	**X**
P: Vehicle cleanliness		**X**
P: Commuting time		**X**
P: Storing personal belongings		**X**
P: Firm branding	**X**	
P: Familiarity with vehicle		**X**

Note: (F) indicates a functional barrier; (P) indicates a perceptual barrier

As part of the online survey, participants were asked to validate the driver-related barriers to the relevance of driver- and assignment-related conditions. The results (Table 4.7) indicated that all previously identified driver-related barriers had a left-skewed distribution well above the scale mean and therefore were highly relevant. The three most important barriers were cleanliness of the driver's cab, assignment of claims, and an easy handover processes.

This section identifies the obstacles to conceptualising a TS concept for road freight transport. The obstacles are classified as functional and perceptual and have varying relevance to the two stakeholder groups under consideration—companies and truck drivers. This work provided the basis for the subsequent implications and solution approaches outlined in Sect. 4.3, aimed at overcoming the barriers.

Section 4.2.2 provides an in-depth examination of the legal implications of direct involvement of truck drivers in the sharing process, as well as the sharing of truck drivers in addition to vehicles.

Table 4.7 Driver-related barriers to truck-sharing

Items ($N = 113$)	*M*	*SD*
1) Cleanliness of vehicles inside	6.25	1.66
2) Cleanliness of vehicles outside	5.71	1.68
3) No extension of commuting time to reach vehicles	5.58	1.81
4) Ability to carry personal belongings in vehicles	5.85	1.87
5) Easy handover process of vehicles	6.12	1.64
6) Familiarity with vehicles	5.60	1.81
7) Allocation of claims with vehicles to the responsible driver	6.13	1.73
8) Plannable working hours for the driving personnel	5.66	1.84

Note: Items were rated on a 7-point Likert scale.

4.2.2 Legal Considerations Regarding Truck Driver Sharing

As outlined in the conceptual fundamentals (see Sect. 2.4.2), the functioning of TS requires a vehicle, load space, and truck driver as prerequisites for providing road freight transport services. TS being designed as an asset sharing model, the vehicle is inherently part of the sharing process. Although a truck driver is a necessary component of the operational process, the truck driver does not have to be included in the sharing arrangement. In the context of car sharing, the focus is exclusively on the asset. However, in view of the shortage of truck drivers, they appear to be a limiting factor in the field of road freight transport. Consequently, the TS can be extended to encompass the truck driver, whereby the vehicle, load space and truck driver are considered unified entities. In contrast to the pure sharing of vehicles, which is essentially a form of rental, labour law considerations also have to be taken into account.

The following considerations are limited to the form of TS in which the driver of the truck is also hired and not just the truck as a vehicle. In this context, the pitfalls that can arise when a transport company uses the concept of TS and what needs to be considered are examined. The main legal basis is the Swiss Federal Act on the Employment Services (AVG) and the Ordinance on the Employment Services (AVV).

Forms of staff leasing

Three types of staff leasing are differentiated: (1) Temporary work (Temporärarbeit, Art. 27 Abs. 2 AVV), (2) Subcontracted work (Leiharbeit, Art. 27 Abs. 3 AVV), and (3) Occasional temporary work (Gelegentliches Überlassen. Art. 27 Abs. 4 AVV). These three forms are each briefly characterised and demarcated below. Subsequently, whether a form of staff leasing requires authorisation, an administrative hurdle, is examined.

Temporary work: The objective of temporary work is to cover demand peaks within the company and to facilitate the deployment of employees in a flexible manner. A temporary employment agency does not engage in direct employment of the workers that it places in assignments. Instead, the business model of a temporary employment agency is based on the provision of workers to a hirer (Bachmann, 2010). In the context of temporary work, an employment contract is typically established between the temporary employment agency and the employee. Its formation is contingent upon the emergence of a specific assignment within a hirer's company. Consequently, the employee receives a payment solely in the event of an assignment being made and for only the duration of the assignment. Notably, the framework employment contract is not, in fact, an employment contract and does not create any obligation to pay a salary. However, the employee is entitled to accept or decline an assignment but has no right to receive an assignment (Greub, 2018) (Fig. 4.6).

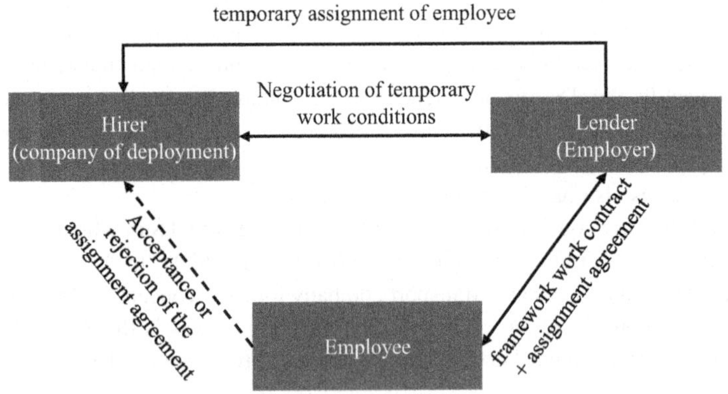

Fig. 4.6 Process of temporary work assignment

Temporary work is not relevant to the concept of TS, because of the nature of its design. Such a model is uncharacteristic and not found in practice among truck drivers. Typically, truck drivers are permanently employed by a company and undertake transport on its behalf. Therefore, the concept of temporary work is not considered further.

Subcontracted work: In contrast to temporary work, in subcontracted work, the employment relationship between the lender and the employee is maintained before or after an assignment, and the employee typically works at the employer's company. Consequently, the employer is obliged to continue to pay the employee's salary even in the absence of an assignment (Looser, 2015). Another distinction is that the employee is obliged to accept an assignment at a host company, the hirer. The lender assumes responsibility for the employee's salary and any associated employment risks at the conclusion of the subcontracting agreement, and this responsibility persists throughout the assignment. However, the authority to issue instructions and the competencies for target and technical instructions transfer to the company with which the employee is deployed. This transfer of authority occurs even in the absence of a contractual relationship between the employee and the hirer (Schweizerischer Bundesrat, 1985) (Fig. 4.7).

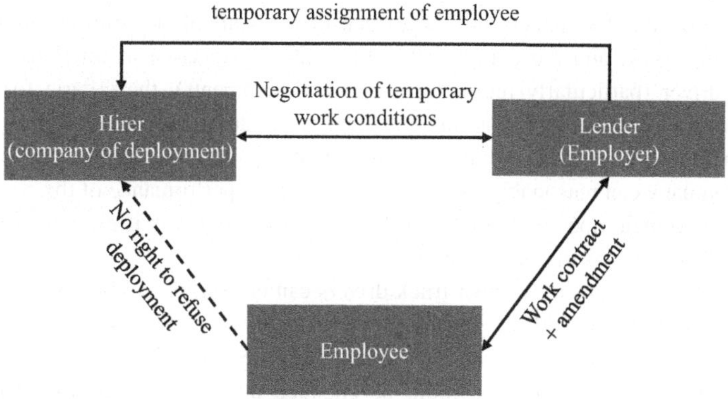

Fig. 4.7 Process of subcontracted work assignment

Occasional temporary work: In this form of temporary employment, employees are made available to a company on a short-term basis without any special planning. This process enables the temporary employment agency to avoid having to lay off employees under economic downturns, for example, after losing an important contract, and can lend employees to companies to bridge gaps until the order situation changes (Seco, 2003). In contrast to the first two forms, occasional temporary work does not require authorisation in Switzerland and is not classified as commercial staff leasing, other than temporary work and subcontratcted work. To be considered occasional temporary work under the Swiss Employment Services Ordinance, the following requirements must be met cumulatively, in accordance with Art. 27 Para. 4 AVV: (1) the purpose of the employment contract between the employer and the employee is that the employee works mainly under the employer's authority; (2) the employee is assigned to a host company only in exceptional cases; and (3) the duration of the employment contract is independent of any assignments with host companies. Non-commercial occasional lending is not discussed in further detail, because the lending of employees in TS is not an exceptional situation but a deliberate strategy applied on a regular basis.

Authorisation duty in staff leasing

In the context of staff leasing, the hirer is typically granted substantial contractual authority to issue instructions. Importantly, staff leasing is a distinct form of contract from other types, such as an order or a contract for services. Consequently, emphasis is placed on the forms of staff leasing rather than order relationships, provided that the employee is deployed in the hiring company and is subject to its authority.

Because the TS concept makes a truck driver available to a hiring company using the truck, and the company has the authority to issue instructions to the truck driver (particularly regarding routes and shipments), the criteria for staff leasing, or more precisely for subcontracted work, are met. The truck driver can reasonably adapt to the working hours of the company utilising the vehicle, and that company can reasonably assume the risk of poor performance of the contract. That the vehicle does not belong to the company using it is irrelevant, because of the fulfilment of the other criteria.

A debate pertains to whether truck drivers can be considered merely operating personnel, which would not fall under the category of staff leasing, rather than orders or contracts for services. For example, in a construction context, both the crane and the crane operator might be engaged through a hiring arrangement. Differentiating between operating personnel and staff leasing requires identification of whether, in addition to target instructions, technical instructions are also provided. In the event that specialised instructions are provided, staff leasing is involved (Bachmann, 2010). Because the hiring company employing the truck driver provides specialised instructions regarding the purpose of the deployment, the location, or the handling of the load, this arrangement constitutes staff leasing and therefore might be subject to the consequences of the mandatory authorisation of staff leasing.

Implications of the authorisation requirement for truck-sharing
As previously stated, TS, including truck drivers, is a form of personnel leasing subject to authorisation. Consequently, users of TS would be required to obtain authorisation to implement the concept. Moreover, the rental company would be obliged to provide a security deposit to cover any potential wage claims. In Switzerland, providing a five-figure sum in Swiss francs as security deposit would be required. The need for authorisation is a substantial obstacle to the sharing of truck drivers, because of the considerable financial burden involved. Even if lending companies oppose being categorised as staff leasing companies, their activities risk being classified as such. No option is available to obtain a ruling or a binding statement of opinion. Consequently, companies tend to obtain licences in advance, in case of doubt (Bachmann, 2010).

Conclusion: Labour law conditions must critically be considered as soon as personnel, typically truck drivers, become part of a TS agreement. In particular, these concerns relate to staff leasing regulations. In the event that staff leasing is used in only isolated instances, and the resulting annual turnover is low (Seco, 2017), the arrangement does not constitute commercial staff leasing. Therefore, the agreements might be assumed to be classified as occasional temporary work, which does not require authorisation. Nevertheless, ambiguity persists regarding the precise criteria, because the exact thresholds remain undetermined and nebulous (Seco, 2017). If staff leasing is classified as commercial, the authorisation requirements apply. Such conditions increase administrative burden and result in capital being tied up in the form of deposits for salaries, thus decreasing the attractiveness of TS from the user's perspective. For TS to be used as a strategic tool by the LSP, it should be used on a regular basis. However, the inclusion of truck drivers poses a barrier. One potential solution is a focused asset sharing approach, wherein only vehicles are part of the sharing.

4.2.3 Investigating the Intention to use Truck-Sharing and Underlying Predictors: Multiple Linear Regression Analysis

To investigate how TS is perceived from the user's viewpoint, and which factors influence the willingness to use TS, this thesis applied the TAM (introduced in Sect. 2.4.4). This model is an effective tool for investigating and explaining the acceptance and utilisation of novel technologies by individuals. The model aids in understanding the factors influencing an individual's willingness to adopt and utilise a novel technology. Given the integral role of digital platforms in SE, technology is a core attribute of TS. The literature suggests that IU is an effective predictor of actual usage behaviour (Davis, 1989). Consequently, despite the currently limited prevalence of TS, the model can be applied in this case.

The literature on the construct of PO suggests that PO is negatively associated with IU, as discussed in Sect. 2.4.4. Moreover, the interviews indicated that the condition of the vehicle is a crucial factor from the user's perspective. In each interview, the significance of vehicle care was emphasised. From the driver's perspective, cleanliness was a notable concern. Additionally, concerns were expressed regarding potential damage to vehicles, in agreement with data from insurers indicating that shared vehicles have relatively high loss ratios. Therefore, ensuring optimal vehicle conditions should coincide with acceptance and openness to sharing trucks. Consequently, the model for hypothesis testing was as illustrated in Fig. 4.8.

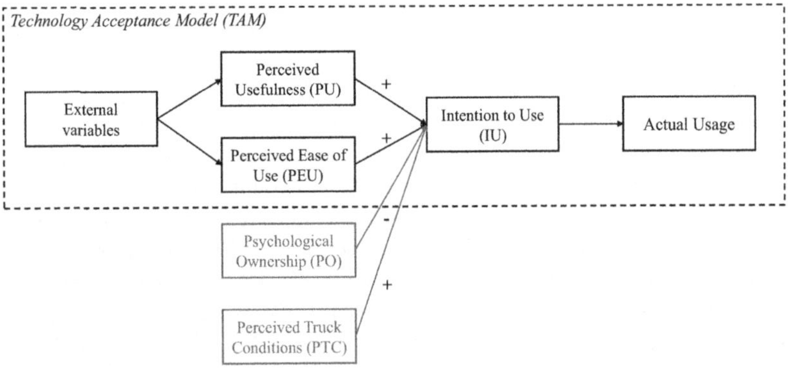

Fig. 4.8 Research model

Items were derived from established scales and adapted to a TS context. Items used to measure IU were adapted from Davis and Venkatesh (1996). PU was adapted from Davis and Venkatesh (1996) and Venkatesh and Davis (2000). PEU was adapted from Davis (1989). The PO and PTC scales were also adapted to align with the research context. For PO, the scale developed by Avey, Avolio, Crossley and Luthans (2009) was used and adapted, limited to the two dimensions of accountability and the sense of belongingness. The scale for PTC was developed anew for this purpose.

Preparatory analyses

Table 4.8 presents the descriptive statistics for the study variables: IU, PU, PEU, PO, and truck conditions. Overall, the IU within the sample demonstrated moderate intensity and inhomogeneity, as indicated by the high standard deviation ($SD > 2$) for the three construct items. The constructs PU and PEU also fell within the moderate range of the scale. The response pattern clearly illustrated considerable divergence in views regarding the utility of TS, ranging from high agreement to strong rejection.

The sample displayed a pronounced tendency towards PO, as evidenced by the left-skewed distribution, particularly regarding truck drivers. Whereas the mean item scores for the entire sample were 5.40 and 5.96, the mean values for the 61 truck drivers were significantly higher, at 6.22 and 6.73. In contrast, the mean values for the two PTC items were below the scale mean. Only a minority of respondents believed that vehicles are kept clean by their users ($M = 2.78$). Furthermore, the response pattern regarding increasing damages due to sharing was even more pessimistic ($M = 2.22$).

Internal consistency

Internal consistency was measured by calculation of Cronbach's alpha for the individual scales. The overall measures for the five constructs are reported in **Table 4.9**. Because all items had an alpha above the standard guideline of 0.70, the scales could be used for analysis with high reliability.

Average variance extracted (AVE) was used as a measure to assess the convergent validity of the scales. An AVE value of 0.5 is considered a good value for adequate convergent validity. The three scales derived from the TAM—IU, PU, and PEU—met this criterion. According to Bagozzi and Yi (1988), the AVE for scales with only two items tends to be unreliable and not very meaningful, because the variance distribution depends heavily on the specific correlations between items. The constructs PO and PTC, which were each measured with only two items, were therefore not subjected to AVE analysis, given that the AVE might have been less stable and meaningful. Instead, the inter-item correlation was considered a reliable indicator of construct reliability. The inter-item correlation of PO was $r = .74$, and the inter-item correlation of PO was $r = .68$. Both constructs thus showed high correlation indicating adequate construct reliability.

Table 4.8 Descriptive statistics of the relevant study variables

Construct	Item	*M*	*SD*
Intention to use (IU)	**IU1**: Provided I have access to truck-sharing, I will take advantage of the offer.	3.50	2.10
	IU2: I would participate in a truck-sharing model if a corresponding offer were available on the market.	3.65	2.15
	IU3: I myself would be open to participating in a truck-sharing concept.	3.52	2.22
Perceived usefulness (PU)	**PU1**: Truck-sharing could be implemented sensibly in my company.	3.20	1.92
	PU2: Truck-sharing brings advantages for transport companies overall.	4.17	1.96
	PU3: The use of truck-sharing increases flexibility for transport companies.	4.48	2.01
Perceived ease of use (PEU)	**PEU1**: For me, the use of truck-sharing in my profession would be easy to implement.	2.92	1.94
	PEU2: Using a 'sharing truck' would not pose any challenges for me in my day-to-day work.	3.18	2.03
	PEU3: To me, truck-sharing seems to be easy to use.	3.21	1.89
Psychological ownership (PO)	**PO1**: The truck I usually drive feels like my own.	5.40	1.95
	PO2: I feel personally responsible for the truck I drive	5.96	1.68
Perceived truck conditions (PTC)	**PTC1**: I believe that shared vehicles are kept clean by users.	2.78	1.76
	PTC2: I assume that less damage to vehicles occurs through truck-sharing.	2.22	1.61

Table 4.9 Construct reliability parameters

Construct	Cronbach's alpha	AVE
Overall ($N = 13$)	0.928	
Intention to use (IU)	0.973	0.925
Perceived usefulness (PU)	0.906	0.768
Perceived ease of use (PEU)	0.867	0.688
Psychological ownership (PO)	0.845	0.300
Perceived truck conditions (TC)	0.807	0.213

AVE = average variance extracted

Construct validity

Factor analysis with oblimin rotation was conducted to ascertain that IU, PU, PEU, PO, and truck conditions were distinct constructs. The suitability of the data for factor analysis was evaluated with the Kaiser-Meyer-Olkin (KMO) measure of sampling adequacy and Bartlett's test for sphericity. The KMO value for the dataset was 0.862, thereby indicating adequate suitability. This finding suggested that the correlations among the variables were sufficiently high; therefore, the data were appropriate for factor analysis. Additionally, Bartlett's test for sphericity was significant ($p < 0.001$), thereby confirming that the correlation matrix significantly differed from an identity matrix, and further supporting the data's suitability for factor analysis.

The results of the factor analysis in Table 4.10 revealed three underlying factors with eigenvalues greater than 1.0, which accounted for 79.5% of the total variance. Three clearly independent factors were confirmed, showing no cross-loadings above 0.3. The only deviating variable was PEU1, with a moderate cross-loading on factor 3 of.356, indicating a good discriminant validity overall. In particular, construct validity confirmation applied to PO and PTC fitting the assumed factor structure. However, items belonging to the TAM deviated from the assumed factor structure. In contrast to the assumed formation of a distinct factor, all items pertaining to the constructs PU and PEU were observed to load collectively with the IU items on a single factor, identified as an overarching technology acceptance factor. Although unexpected, this pattern is not unusual in studies applying the TAM, because the constructs PU, PEU, and IU exhibit moderate to strong correlations. In a quantitative meta-analytical review of 569 findings derived from 95 TAM studies, Yousafzai, Foxall, and Pallister (2007) have revealed significant diversity in the relationships reported within the extant literature. The data indicated that certain correlates exhibited considerable variability across the reported values. Moreover, the authors identified significant correlations between constructs, as follows: $r = .91$ between PU and IU; $r = .78$ between PEU and IU; and $r = .81$ between PU and PU. The correlations among these three constructs in the sample were in the upper range and approached the highest correlations observed in other studies, as demonstrated in **Table 4.11**. These high correlations among the constructs suggested the presence of a higher-order factor potentially representing general technology acceptance related to TS.

Table 4.10 Factor loadings of items

Construct	Item	FA1	FA2	FA3
Intention to use (IU)	IU1	**.752**	−.157	.242
	IU2	**.807**	−.210	.197
	IU3	**.810**	−.198	.170
Perceived usefulness (PU)	PU1	**.808**	−.079	.139
	PU2	**.975**	.083	−.201
	PU3	**.938**	.119	−.211
Perceived ease of use (PEU)	PEU1	**.646**	−.036	.356
	PEU2	**.582**	.126	.149
	PEU3	**.712**	.201	.113
Psychological ownership (PO)	PO1	−.007	**.823**	.075
	PO2	.024	**.821**	.051
Perceived truck conditions (PTC)	TC1	.047	.060	**.799**
	TC2	.122	.123	**.677**
Eigenvalue		7.426	1.833	1.073
Cumulative percentage of variance		57.121	71.217	79.474

Note: Pattern matrix of the principal axis factor analysis, oblimin-rotated with Kaiser extraction. The highest factor loadings per item are indicated in bold.

Table 4.11 Correlations among study variables

Construct	1.	2.	3.	4.	5.
1. Intention to use (IU)	–				
2. Perceived usefulness (PU)	.795**	–			
3. Perceived ease of use (PEU)	.749**	.807**	–		
4. Psychological ownership (PO)	.026	.170	.212*	–	
5. Perceived truck conditions (PTC)	.566**	.482**	.623**	.186*	–

** Correlation significant at 0.01. * Correlation significant at 0.05.

Multiple linear regression analysis

For testing the theoretical model with the hypotheses derived in Fig. 4.8, a multiple linear regression analysis was conducted. Control variables were not included in the exploratory stage, according to the recommendations of Bernerth and Aguinis (2016) regarding the use of control variables in organisational research. In the current state of research, no control variables were included in the analyses without initial empirical evidence on TS acceptance in practice, in view of the novelty of the construct under investigation.

In accordance with the TAM, PU and PEU were hypothesised to be positively associated with the IU TS. Furthermore, PO was hypothesised to be is negatively associated with IU and—on the basis of explorative interview findings—PTC was hypothesised to be positively associated with IU. The results of the regression analysis are shown in Table 4.12.

First, for the relationship with IU, positive relationships with PU ($\beta = .558$, $p = .000$**) and PEU ($\beta = .210$, $p = .037$*) were found. The results substantiated existing findings based on the TAM, and suggesting that ease of use and usefulness are key predictors of IU. Furthermore, the results indicated that PU was a dominant predictor.

Second, the hypothesis of a negative correlation between IU and PO was confirmed ($\beta = -.149$, $p = .007$*), thus indicating that the IU TS decreases with increasing attachment to a particular vehicle. In contrast, a lower level of driver's attachment to the truck would result in higher acceptance of TS.

Third, perceived truck conditions were confirmed to be positively associated with IU (ß = .193, $p = .005$*). Among respondents, greater conviction that users ensure vehicle cleanliness and damage prevention through TS was associated with greater willingness to engage with a TS concept.

Table 4.12 Results of linear multiple regression analysis

Construct	*B*	*SE*	*ß*	*T*	*p*
Constant	1.723	1.293		1.332	0.186
PU	0.652	0.104	**0.558**	6.281	0.000**
PEU	0.255	0.121	**0.210**	2.108	0.037*
PO	−0.277	0.100	**−0.149**	−2.767	0.007*
PTC	0.395	0.138	**0.193**	2.874	0.005*

Note: * $p < .05$. ** $p < .01$.

Conclusion: The quantitative analysis confirmed that PU and PEU are significant predictors of IU, and that PU has a strong influence. PO negatively influenced IU, whereas positive perceptions of truck conditions positively influenced IU. Overall, the tested model was confirmed. Beyond demonstrating the applicability of TAM in a TS context, the findings highlighted that firmer belief in the care and condition of shared trucks positively correlated with intention to use TS. Additionally, the importance of addressing PO in promoting the adoption of TS was highlighted.

The degree of openness to sharing observed in the sample was rated as moderate overall. During the interviews and subsequent quantitative analysis, several factors differentiating behavioural intention to share trucks were identified. The following section presents typical truck deployment and truck driver profiles characterised by willingness to share, as well as links to use cases identified in Sect. 4.1.3.

4.2.4 Truck Driver and Deployment Profiles: Deriving the Sharing Persona

Driver-centred determinants of TS usage intention: For identifying the characteristics of a typical sharing driver, several criteria were identified that distinguished truck drivers in the context of their workplace. These criteria included sociodemographic variables, such as age and gender, and driver-related distinguishing features, such as the frequency of overnight stays in trucks, transport distances performed, the degree of personalisation of trucks, and the truck occupancy profiles, given that trucks may be used by one truck driver exclusively, in small teams, or in flexible, changing settings (Fig. 4.9).

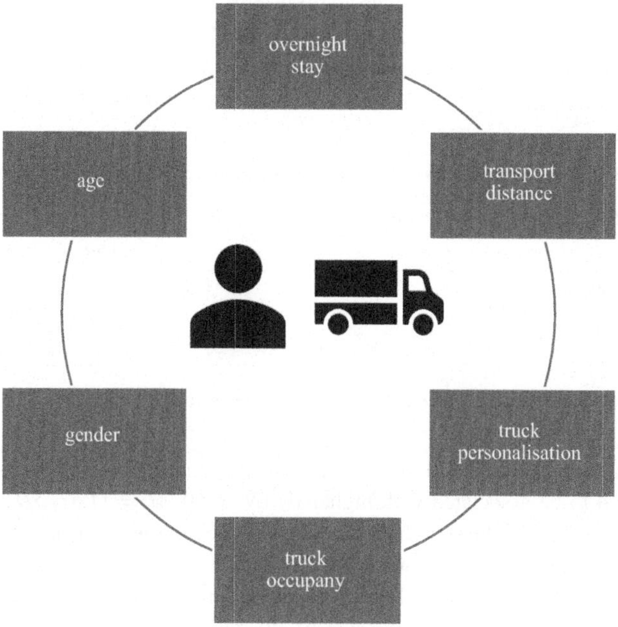

Fig. 4.9 Criteria for prototyping a truck-sharing profile

Overnight stay and transport distance: Truck drivers often transport freight over very long distances of several hundreds or thousands of kilometres, and sometimes across national borders. Many truck drivers are on the road for several days or weeks. This extends the significance of the truck beyond the vehicle itself. Trucks may serve as mobile homes with provisions, personal belongings, and an integrated cabin for sleeping. In the sample, 46 of the 63 truck drivers stated that they spent at least one night per month in their truck (Table 4.13). More than half regularly spent the night in their truck. In contrast, only 17 truck drivers stated that they never spent the night in their truck. Truck drivers who never slept in their trucks had the highest IU TS ($M = 14.88$, $SD = 5.87$). In contrast, the level of agreement was significantly lower in the other three groups. Truck drivers were considerably less willing to share the driver's cabin with other truck drivers if the truck was also used for sleeping purposes, regardless of how often they sleep in their truck.

The results were consistent with the transport distance profiles. Truck drivers driving primarily long-distance routes ($N = 43$, $M = 11.63$; $SD = 5.043$) considered TS to be less useful on average than truck drivers driving primarily short- and medium-distance routes ($N = 20$, $M = 13.00$; $SD = 5.729$).

Table 4.13 Intention to use descriptives among truck drivers by overnight stay frequency

Overnight stay frequency per month ($N = 63$)	N	M	SD
a) Never	17	14.88	5.87
b) 1–2 nights	3	8.00	8.66
c) 3–5 nights	10	8.36	5.51
d) >5 nights	33	9.42	6.22

Note: IU construct score is calculated as the sum of the three IU items

Truck personalisation: In terms of truck personalisation, standardised vehicles tended to be associated with higher IU ($N = 50$, $M = 11.57$, $SD = 5.845$). When truck drivers drove vehicles configured according to their own wishes, the IU the vehicle was somewhat lower ($N = 63$, $M = 10.41$, $SD = 6.334$). The results indicated an initial trend. However, in light of their high variation, the results were not statistically significant. Moreover, no causality could be assumed. Further investigation of the extent to which the groups differ would be worthwhile. Does standardisation of vehicles favour an increased tendency to share? The interviews indicated that SMEs were strongly committed to their truck drivers and intensively advertised to have sufficient driving personnel. The offer of vehicle customisation was often provided as an incentive.

Truck occupany: Compared with the status quo, TS was associated with a change in how vehicles are used. The current usage behaviour was identified to influence truck drivers' attitudes towards TS. As shown in Table 4.14 companies with exclusive vehicle use were less open to sharing than those with use of vehicles by several truck drivers. In practice, models are widespread in which a truck is permanently assigned to a small group of truck drivers who know one another. Very flexible models, in which no fixed assignments are made, similarly to a rental vehicle scenario, are rare. Respondents familiar with flexible usage concepts from their everyday working lives showed the highest levels of agreement in the survey. In-depth truck driver interviews suggested that regular vehicle changes were associated with lower attachment to the truck. In that scenario, truck drivers are unable to leave personal items in the truck. The activity in such a model is already much closer to potential TS, so the change in behaviour is less pronounced.

Table 4.14 Intention to use descriptives among truck drivers by truck occupany

Truck occupancy ($N = 113$)	N	M	SD
a) Exclusive use: trucks are allocated to individual people (if truck driver X is sick, the vehicle remains in the yard)	55	9.64	6.18
b) With one colleague: trucks are shared by colleagues	26	11.88	6.98
c) Flexible: vehicles are regularly changed with no fixed allocation.	9	14.11	3.41
d) Other use	23	11.43	5.70

Note: IU construct scores were calculated as the sum of the three IU items.

Age: The effects of age were investigated by examination of socio-demographic variables. A greater tendency towards sharing services was observed among younger than older individuals. In addition to social factors such as lifestyle, other mobility needs, and financial motives, higher affinity for technology was a significant driver of this phenomenon. Younger generations are more inclined than older generations to embrace technology and to utilise mobile applications and digital platforms, which are often essential for the reservation and administration of car-sharing services. The effects of age on usage patterns was investigated, but no correlation was identified ($\beta = -.03$, $p = .595$). The younger respondents did not demonstrate greater openness than the older respondents. Therefore, age was not identified as a predictor.

Gender: In terms of gender, no reliable quantitative evaluations are provided. In the sample, 7 of 63 truck drivers were women, corresponding to a share of 11%, a percentage notably higher than the industry-wide share of women of only 2.2% (BALM, 2024). Because of the small sample size, further quantitative evaluations were therefore not performed. The consideration of gender-specific topics is becoming increasingly important in road freight transport, in light of the truck driver shortage. Whereas the number of new-entry truck drivers is declining overall, the number of women in the profession is continually increasing, from a low level (BALM, 2024). From 2013 to 2022, the number of female truck drivers in Germany increased by 40%, from 1.6% to 2.2%. This trend is likely to continue in the future, given that the proportion of women in newly concluded training contracts has exceeded the 10% threshold.

Factors that promote and hinder sharing

On the basis of the determinants, profiles associated with an open or rejecting attitude were identified. The beneficial and hindering factors are used in this section to outline the dichotomy between individuals who are open to embracing TS, referred to as *sharers*, and those who resist this approach, referred to as *opponents*. The insights gained from this exploration contribute to a broader understanding of TS barriers and drivers, thereby enabling practitioners to differentiate among their own organisation's truck drivers, and determine truck deployment according to TS acceptance potential. In Table 4.15 the prototypical personas of the *sharer* and the *opponent* are contrasted.

Table 4.15 Contrasting the sharer and opponent

Dimension	The Sharer	The Opponent
Employment status	*Employee*: Employed truck drivers do not own their vehicles. Willingness to share vehicles is high, because the truck driver is not tied to owning a truck.	*Sole entrepreneur*: Self-employed truck drivers tend to be sceptical about sharing. These truck drivers usually operate exactly one self-owned truck and often work as subcontractors for other LSPs.
Transport distance	Distances comprise short to mid-range tours, often with a constant rhythm and fixed transport lanes.	Distances comprise long-distance transport across Europe and are associated with regular overnight stays in trucks.
Overnight stay	The driver returns home every day and never sleeps in the truck.	The driver sleeps in the truck regularly.
Truck personalisation	The truck is not personalised at all; the vehicle is interchangeable from the driver's perspective.	The truck is individually configured truck according to truck driver preferences; this configuration is associated with a high sense of truck ownership.
Truck occupancy	The truck driver is not vehicle-bound and switches vehicles with colleagues regularly; thus, the sense of truck ownership toward one specific vehicle is not pronounced.	The truck driver is accustomed to being the sole operator of the vehicle. If the truck driver cannot work, no other person is assigned to take over the vehicle.

(continued)

Table 4.15 (continued)

Dimension	The Sharer	The Opponent
Digital affinity	The driver is open to new technologies and regularly uses smartphone apps.	The driver has limited experience with smartphones and new technologies in general, which are viewed with scepticism.

The presentation of the contrasting sharer and opponent personas completes the description of the results in Sects. 4.1 and 4.2, which were geared towards RQ1 and RQ2. The following Sect. 4.3 discusses the results and implications associated with application-oriented recommendations focusing on RQ3.

4.3 Implications and Recommendations

In projecting the current challenges in road freight transport outlined in Sect. 1.1 into the future, prior findings revealed four underlying macro trends supporting the adoption of TS in road freight transport over the next decade: the continuing *engine transition* towards zero-emission vehicles accompanied by a shifting cost structure of truck operations; the increasing *truck driver shortage* across Europe, thus mounting pressure to increase the utilisation of existing capacity; the adoption of *autonomous truck operations* affecting truck deployment profiles and the chance of extending utilisation intensity to round-the-clock operation; and *infrastructural constraints* paired with expected freight volume growth, which increase the need for more efficient utilisation of load and parking space to relieve the road infrastructure.

The confluence of these four factors provides a favourable context for fostering TS. However, no driver individually or in combination is sufficient to drive the implementation of TS in road freight transport. From a normative perspective, several operational barriers must be addressed to promote TS. Some of these barriers, such as the need for truck drivers to exchange vehicles, are not open to influence, because they are contextual factors.

In addition, multiple barriers that can be actively addressed were identified. The following paragraphs address these barriers, to identify possible solutions for designing a TS concept.

Stakeholder-integrated coping with multiple barriers

The analysis of barriers to TS in road freight transport in Sect. 4.2.1 revealed the existence of multiple barriers. The barriers were divided into two main categories: functional barriers, which are necessary conditions for TS to be feasible and efficient, and perceptual barriers, which reflect individual attitudes towards TS. Functional barriers concern primarily companies and include issues such as supply demand matching, legal compliance, and vehicle handover processes. Perceptual barriers affect primarily truck drivers and relate to concerns regarding vehicle cleanliness, commuting time, and vehicle familiarity. The application of the TAM to investigate the IU in Sect. 4.2.3 also confirmed PU as a strong predictor of TS use. Notably, actual usefulness is distinct from PU, because the two constructs do not always perfectly align. The foundational TAM model by Davis (1989) suggests that, even if a system is objectively useful, the user's perception determines adoption. For instance, a technology might be objectively effective, but if users do not perceive it as useful, they might not adopt it.

These findings have implications in the relevance and handling of barriers. First, both functional and perceptual barriers are relevant to the IU, and have been demonstrated to influence PU. Second, overcoming these barriers requires strategies at both the organisational and individual levels, with actor-specific approaches. When designing TS, explicitly considering the needs of truck drivers is essential to address the perceptual level. Whereas the decision to use TS is ultimately a business decision, TS implementation requires acceptance from truck drivers. The objective is to align the added value for the company with the needs of the employees.

Conceptual implications for truck-sharing design

Analysis of the interviews consistently indicated the enabling role of a digital platform for the organisation of exchange. A digital platform is considered essential and central to organising TS, by facilitating the seamless coordination and management of available freight vehicles. Such platforms enable efficient vehicle allocation and ensure compliance with logistical and legal requirements by providing a real-time interface for matching supply with demand, and further support communication among involved stakeholders, thereby fostering transparency and trust. Moreover, the platform may incorporate advanced features such as dynamic pricing, tracking, and analytics, thereby further enhancing TS initiatives' operational efficiency and scalability.

An algorithm-based matching of supply and demand is performed, taking into account all requirements and restrictions. Given the high volume and variety of data to be integrated, this matching process would be impractical to perform manually. To ensure efficiency in the organisation of the TS platform and in the practical process, the platform must be designed specifically for this application. Solution approaches could be derived for this purpose.

Trust is crucial for in the success of TS systems, because it underpins the willingness of companies and individuals to engage in cooperating and sharing assets that are critical to their operations (e.g., Pomponi et al., 2015; Daudi, 2018). Trust includes several dimensions, such as trust in the conditions of vehicles, trust in the reliability of the system, trust in the platform, or trust in sharing participants Without trust in the system, the platform, and other participants, companies and individuals are unlikely to engage in sharing, thereby limiting the potential benefits of the model. Given this background, recommendations for the design are derived and presented:

Keyless and contactless vehicle takeover: The unattractiveness of ultra-short-term truck rental in the classic rental market is due to administrative costs outweighing the revenue generated when trucks are rented for only several days or hours. An app-based, keyless handover could markedly reduce these costs and make short-term truck rental more viable. This facilitation of sharing by the hour or day is crucial for TS. By minimising administrative effort, the hurdle of ultra-short-term rental is removed, the economic potential of TS is maximised, and the functional barrier of the vehicle handover process is addressed.

The efficacy of such solutions has been demonstrated through successful trials in the context of car sharing, in which they have become standard practice. After the reservation process is completed, users are granted access to the vehicle via an app, after which they may proceed to unlock it. Such solutions are not exclusive to business-to-consumer sharing, as evidenced by the utilisation of peer-to-peer car sharing models. Private car owners are provided with a telematics device allowing recording of vehicle data. This device is connected to the vehicle's on-board network, thus enabling communication between the car and an app. The user downloads an app that communicates with the vehicle's telematics system and enables users to book, locate, and unlock the car. This principle could applied to trucks.

Image/video recording as documentary evidence: For truck drivers, the cleanliness of the vehicle is a relevant consideration. Ensuring that rented vehicles are returned undamaged is in the company's interest. Both aspects were identified as potential barriers to TS success. External users are generally assumed to treat rented vehicles less carefully, thus potentially increasing cleanliness and damage issues. One potential solution might be the use of image and photo evidence. After each use, the user could be required to document the condition of the vehicle, both internally and externally. To this end, an app solution could be enhanced with functionality to upload images and videos associated with the rental process. In particular, the truck driver would be required to provide proof of the vehicle's interior and exterior condition. Truck drivers are recommended to walk around the vehicle and record it from all sides.

The use of artificial intelligence-based image recognition enables the automated comparison of images and videos to verify the condition of a vehicle after it has been returned, by comparing the current condition with the condition before use. This process minimises the need for time-consuming verification. Only in the event of an incident would a service employee be called to examine the case in detail.

The process also helps track damage occurrences and identify causes, thereby allowing the system to encourage careful handling and attribute incidents to their causes.

Processing of financial transactions: The research findings indicated that a peer-to-peer TS platform could not only facilitate interactions between vehicle owners and renters, but also handle financial transactions. First, the platform's management of financial transactions would facilitate the establishment of trust between users. By acting as an impartial intermediary, and ensuring the secure and accurate processing of payments, the platform would enforce confidence in the entire process. Trust is a critical factor in the acceptance and use of peer-to-peer services, as evidenced by the TS industry. Second, the integration of payment processing would help mitigate risks to all parties involved. The platform must critically ensure that the vehicle owner receives payment before the vehicle is handed over and must simultaneously protect the renter in the event of problems with the vehicle. Mitigating risk is a crucial aspect of ensuring the safety and reliability of the platform. Third, centralised payment processing offers users a convenient solution. By assuming responsibility for all administrative tasks related to payment, the platform would streamline the process for both parties and enhance the user experience, thereby strengthening user loyalty and increasing the attractiveness of the platform.

Overall, the integration of financial transactions strengthens trust, efficiency, and security within the platform, and is crucial for its long-term success. These recommendations should therefore be considered strategic measures to optimise the platform.

In this context, the derivation of recommendations for a pricing system is intentionally excluded from the scope of this thesis and is to be addressed in subsequent business model development.

Integrated insurance solution: In order to overcome the insurance-related obstacles, it is recommended that a standardised integrated vehicle insurance solution be implemented by the platform, for the benefit of both parties to the contract, which will take effect in the event of sharing. Such insurance solutions have become a standard feature of station-based and free-floating car sharing for fleets. Furthermore, the integrated approach is also pertinent to peer-to-peer car sharing, given that the rental of vehicles owned by transport companies typically does not encompass commercial rentals. Integrated coverage through the sharing platform allows users to focus on their core competencies, eliminating the administrative burden of vehicle insurance. It is beneficial that no modifications are required to the existing insurance policy.

Concerns regarding an increase in vehicle damage due to sharing can be addressed by exploring fully comprehensive insurance options with capped user liability. This has the advantage for the owner of not bearing the costs of damage. Users have the knowledge that they can plan up to which amount they are liable. The risk of increased damage is assumed by the insurance company, which charges a premium for this.

Fuel card-based refuelling process: Ensuring that fuel and energy costs are transparent and allocated in accordance with the polluter-pays principle is critical to facilitate user acceptance. Therefore, vehicles are recommended to be equipped with platform-managed fuel cards that are billed directly via the platform. To consolidate the payment flows, combining the costs of fuel and charging with those of vehicle rental or leasing is recommended.

Furthermore, making vehicles available for unrestricted use immediately upon collection is in the interest of the users. To ensure that the vehicles can be scheduled and utilised as intended, a minimum fuel or charge level must be maintained at the time of return. This aspect could be monitored centrally by the platform through the use of the telematics system installed in the vehicle. If this aspect proves challenging in practice, a financial penalty could be imposed as part of a potential penalty system for non-compliance.

Toll accounting and liability: Tolls must paid without delay. In practice, fleet operators typically encounter liquidity drain at a later stage, as fuel card providers are positioned between the operator and the toll settlement. These providers offer a payment guarantee to the toll operator, assuming liability for the toll. To mitigate risk, these companies conduct credit checks in advance. If a company is unable to identify a fuel card provider because of an unfavourable credit rating, it must replenish its credit and maintain liquidity. If the credit is depleted, the trucks must vacate the toll road at the subsequent exit point. For rental companies, the vehicles are registered in the name of the rental company, which is held liable for the toll in the event of a default in payment and outstanding toll costs. In the event of a user failing to settle toll charges in a timely manner during the rental period, the responsibility and thus the liability might fall upon the rental company, thereby posing a financial risk to the rental company. The rental company might be required to pay any outstanding charges regardless of whether the user is responsible for them. In the case of TS, the entity to which the vehicle is registered is held liable for the toll payment. In the case of long-term rental, the lessor has the option of registering the vehicle with the toll provider in the name of the lessee, thereby transferring liability to the lessee as the user. This option is not feasible in the context of TS, given the relatively short rental period.

A credit check process is considered an effective method for identifying users unable to meet their financial obligations. As part of the registration process, an automatic credit check could be initiated. Moreover, contractual agreements could be used to ensure that the rental company is not held liable for any default in payment. The rental agreement might include clauses requiring the renter to utilise the on-board unit[2] correctly during the rental period and to ensure payment of all toll charges incurred. Additionally, an obligation to provide a means of payment that debits the toll charges directly could be agreed upon. Thus, the platform could facilitate security and trust.

Vehicle quality check: Truck condition was found to be both a predictor of, and an obstacle to, IU TS. Cleanliness and damage were described as anticipated outcomes. Whereas truck condition could be monitored in terms of appearance through a photograph- and video-based inspection system, technical conditions are an additional dimension from a security perspective.

However, these conditions cannot be recognised at a glance. To ensure technical functionality and thus support trust in the vehicles provided, a quality check should be introduced. The platform could prescribe mandatory regular maintenance intervals, during which vehicles are checked and maintained by certified workshops. This process could be monitored via a digital maintenance logbook on the platform. In addition to other ongoing measures such as user feedback, a gatekeeper function should be introduced at the outset. Before being authorised, all vehicles should undergo inspection by a service centre, so that only vehicles that are reliably free of technical defects are authorised for operation. The pre-check would ensure the quality of the vehicles.

---

[2] An on-board unit is a device installed in a vehicle to enable automatic billing in a toll collection or toll system. The use of an on-board unit eliminates long waiting times at toll stations and the need for truck drivers to make complicated cash payments. Tolls are automatically recorded and billed via the electronic vehicle devices when the vehicle passes through the toll station.

Conclusion: The described bundle of measures, which are platform-related and closely aligned with the operational sharing process, emphasises the dominant role of the digital platform, whose functionality extends far beyond mere brokering of matches. The list of platform-related functionalities is not exhaustive, and other functionalities could potentially be implemented in the operation. For example, an optional pick-up and delivery service for vehicles could facilitate the organisation of geographically peripheral matches. These and other services should be tested continually in the course of a pilot operation.

Discussion 5

5.1 Summary of Research Findings

The objective of this thesis was to investigate the potential of TS as a solution to enhance efficiency in road freight transport, addressing challenges such as growing freight volumes, infrastructure constraints, low profit margins, and truck driver shortages in Europe. Aims included elaborating on the concept of TS, identifying practical use cases, and examining the drivers of, and barriers to, TS implementation. Through an exploratory mixed-method design involving qualitative interviews and surveys, both theoretical insights and practical recommendations were provided to support adoption of TS in the logistics industry.

The thesis was based on three key RQs. First, it assessed the potential uses of TS in road freight transport and how this approach might contribute to greater industry efficiency. Second, it examined the main factors influencing the adoption of TS, identifying factors that facilitate and hinder implementation. Finally, the research examined practical implications and made recommendations to support the successful integration of TS into existing transport operations, to provide actionable insights for industry stakeholders.

A mixed methods research design was used to guide the empirical investigation, combining qualitative and quantitative data collection in an exploratory sequential mixed methods framework. Semi-structured interviews, case studies, and an online survey were used as core methods across the RQs. The research framework integrated these methods into a design blending qualitative and quantitative phases, using backcasting for future-oriented recommendations while ensuring practical relevance through an engaged scholarship approach.

L. Häberle, *Truck-Sharing in Road Freight Transport*, Supply Chain Management, https://doi.org/10.1007/978-3-658-48754-6_5

Potential and use cases

Status quo: Given the limited availability of data, case studies were used to analyse current fleet utilisation, and establish a foundation for potential and use case derivation. The case studies demonstrated that the companies under review had higher utilisation rates than those described in the literature. The case studies confirmed the plausibility of the utilisation factors reported in the literature (e.g., in Pahlén & Börjesson, 2012, who determined utilisation rates of 61% by volume and 64% by weight). However, these theoretically determined values were found to have only limited significance in quantifying the discrepancy between the effective utilisation and the actual maximum utilisation of a truck in operation. A truck may be fully loaded at values below 100%, depending on the transported goods, when either the load space is at maximum volume or the weight limit has been reached. Consequently, two-dimensional utilisation measures that simultaneously consider volume and weight yield results closer to reality. The utilisation rates of 80% to more than 90% determined in the case studies provided a more accurate representation of the actual utilisation of vehicles in road freight transport. In-depth interviews suggested that these values are also representative of the industry. Simultaneously, the results indicated that improvements are possible in all efficiency dimensions examined, to achieve less downtime, fewer empty kilometres, and higher load space utilisation.

Potential: The potential of TS is perceived primarily in terms of enhanced flexibility and diminished downtime. The two primary freight transport efficiency measures, utilisation rate and empty kilometres, are indirectly influenced by changes in the deployment profile, driven by increased operational flexibility. Providing transport companies with an additional flexibility-enhancing tool for fleet management and truck deployment planning enables improvement in primary efficiency measures beyond the status quo, through the identification of superior use cases. Expert estimates indicated that the enhanced efficiency within the system has the potential to slightly decrease the overall number of trucks on roads through the implementation of TS, with an estimated average effect of 2.6%.

Use cases: Several use cases were identified to achieve the two main underlying aims of reducing downtime and increasing flexibility:

1. *Balancing transport demand and fleet capacity*: From an efficiency standpoint, TS addresses the primary pressure point by improving the ability to balance out fluctuations. These fluctuations can be either cyclical, manifesting as seasonality, or unsystematic, driven by economic cycles. Given the limited ability

to scale transport capacities up and down in the short term, TS offers a means to enhance efficiency in road freight transport.

2. *Vehicle downtimes*: Vehicles need not always be shared for entire days or weeks. A centralised TS platform could be used to determine compatible usage periods across companies, combining idle times with transport requirements and thus enabling multi-shift operation. This design would allow use intensity to be increased across company boundaries, because the platform could be used to identify periods when vehicles are not required for transport.
3. *Alternative drive systems*: Closely associated with this is the promotion of the switch to alternative drive systems in road freight transport. TS provides an opportunity to support the market penetration of trucks with alternative drive systems, particularly in reducing financial barriers to smaller transport companies. These firms frequently lack the resources to invest in costly new technologies such as electric or hydrogen-powered trucks. By utilising TS platforms, businesses could gain access to these vehicles on a rental basis, thereby avoiding the high upfront costs of ownership. This approach would allow companies to test alternative drive systems in real-world conditions, such as urban deliveries in low-emission zones, thereby gaining valuable experience with the technology before making substantial investments. This ability would not only make the technology more accessible but also accelerate its adoption in the market.
4. *Financial motivation*: From a financial standpoint, the use of TS is associated with a decline in fixed costs for transportation enterprises, as the number of vehicles in the fleet decreases. Those who temporarily rent out their vehicles also gain access to new revenue streams.
5. *Specialised vehicle access*: TS provides fleet operators with access to specialised equipment, thus enabling them to fulfil a range of transport orders without a need to invest in expensive, dedicated vehicles such as refrigerated units or tankers. Regardless of company size, in some instances, the financial investment or consistent demand might not justify ownership of specialised vehicles. TS provides a flexible, cost-effective solution for renting such vehicles as needed. For instance, a general cargo fleet could rent a refrigerated truck for perishable goods and avoid the cost of underutilised equipment. Specialised truck owners benefit from increased utilisation and additional revenue by renting out their vehicles, thus helping offset ownership and maintenance costs.

Drivers of, and barriers to, truck-sharing

The aim of RQ2 was to identify drivers of, and barriers to, TS in road freight transport. First, consolidated drivers are outlined; subsequently, barriers are contrasted.

Supportive macro trends: The convergence of four macro trends provides a supportive foundation for the adoption of TS in road freight transport in the coming decade. First, the ongoing transition towards zero-emission vehicles will alter the cost structure of truck operations, thereby making shared usage models more economically attractive. Second, the growing truck driver shortage across Europe is intensifying the need to maximise the utilisation of existing capacity, and is causing companies to consider more efficient solutions, such as TS. Third, the rise of autonomous truck operations presents an opportunity to extend vehicle utilisation to round-the-clock operations, and consequently to further enhance the temporal potential for sharing trucks. Finally, infrastructural constraints, coupled with expected freight volume growth, underscore the need for more efficient use of load and parking space, thus making TS a compelling solution to relieve pressure on road infrastructure. Together, these factors create a context conducive to the adoption of TS. However, although these trends provide a strong foundation, each is not sufficient on its own to ensure the widespread implementation of TS in road freight transport.

Truck-sharing and technology acceptance: The TAM was used as a theoretical model to investigate the IU TS and its underlying predictors.

As revealed by the multiple linear regression analysis, the PU and PEU were identified as predictors of behavioural intention to use TS. PU has a particularly strong influence; consequently, when users see clear benefits and find the platform easy to use, they are more likely to adopt TS. In contrast, PO negatively affects the IU TS; therefore, strong personal attachment to a truck decreases the likelihood of participation in sharing schemes. The high vehicle-related PO in the sample indicated strong links between truck drivers and transport companies to the vehicles that they use. These findings fed into the implications and recommendations for action. Additionally, the PTC for shared trucks, particularly regarding cleanliness and damage prevention, positively influenced the intention to engage with TS. These factors collectively underscore the importance of demonstrating the advantages, ease, and maintained condition of shared trucks to encourage TS adoption.

Intention to use truck-sharing: The approval ratings for the IU TS were at the middle of the scale. Neither pronounced euphoria nor fundamental rejection of TS weas observed. This result was not unexpected, given that the idea of TS remains abstract and appeared far removed from everyday life for the respondents.

This finding also reflected the existence of numerous recognised barriers, which are viewed with respect to the status quo.

The sharer versus the opponent: The thesis identified tendencies in truck driver profiles influencing the IU TS, focusing on socio-demographic variables, work-related characteristics, and attitudes toward TS. Key determinants included the following:

- *Overnight stays and transport distance*: Truck drivers who did not spend nights in their trucks and those who drove shorter distances showed a higher IU TS, whereas truck drivers who regularly slept in their trucks or covered long distances were less willing to share, because of the truck's role as a personal space and home.
- *Truck personalisation*: Truck drivers using standard, non-personalised trucks were more open to TS than those with personalised vehicles. The attachment to customised trucks decreased willingness to share, although the results were not statistically significant.
- *Truck occupancy*: Truck drivers accustomed to flexible vehicle usage, such as regularly switching trucks, were more inclined towards TS. In contrast, those with exclusive vehicle use had lower openness, because of a strong sense of ownership and personal attachment to their trucks.

Two distinct personas representing contrasting tendencies emerged from the analysis:

- *The Sharer*: Typically, an employee with no personal truck ownership, engaged in short-to-mid-range transport, not spending nights in the truck, open to flexible vehicle use, and comfortable with digital technologies.
- *The Opponent*: often a self-employed truck owner involved in long-distance transport with regular overnight stays, using a personalised truck, resistant to sharing due to strong vehicle ties, and generally less familiar with digital tools

These profiles highlight the tendencies observed to promote or hinder TS adoption, thus providing insights for organisations to tailor TS strategies according to truck driver characteristics.

Functional and perceptual barriers: The analysis identified a range of barriers to implementing TS that were classified into two main categories: functional and perceptual barriers.

1. *Functional barriers:* These are prerequisites that must be met for TS to be feasible and efficient. They include:
   - ➢ Supply-Demand Match: Aligning vehicle availability with demand
   - ➢ Vehicle positioning: ensuring minimal empty kilometres by properly positioning vehicles
   - ➢ Vehicle feasibility: meeting specific functional requirements according to the type of freight
   - ➢ Refuelling: managing fuel or battery levels and cost allocation
   - ➢ Labour law compliance: addressing legal issues related to including truck drivers in TS
   - ➢ Vehicle insurance: managing increased risk and costs of sharing
   - ➢ Toll accounting and liability: mitigating financial risks related to toll payments
   - ➢ Claims allocation: assigning damage and other claims appropriately
   - ➢ Vehicle handover process: reducing inefficiencies in the handover process, particularly for short-term rentals
2. *Perceptual barriers:* These barriers relate to individual attitudes and perceptions towards TS, particularly among truck drivers:
   - ➢ Vehicle cleanliness: concerns about cleanliness not meeting personal standards
   - ➢ Commuting time: avoiding increased commuting time to access shared vehicles
   - ➢ Storing personal belongings: discomfort with frequently changing trucks and effects on personal space
   - ➢ Firm branding: issues with driving vehicles branded by other companies, particularly competitors
   - ➢ Familiarity with vehicles: challenges in adapting to different trucks, potentially affecting safety and comfort

Legal considerations regarding truck driver inclusion: The provision of freight transport services requires the conjunctive interplay of truck, trailer, load space and truck driver, all of which must coincide. If only one of these production factors is not available, the service provision is disrupted. Truck drivers are therefore a relevant bottleneck factor, also for TS. The greater the shortage of truck drivers in the industry, the more the focus shifts to the sharing of entire capacities, including truck drivers besides physical assets. In order for LSPs to apply TS as a strategic and value-generating tool, it should be applied on a regular basis. However, including truck drivers represents a legal impediment that gives rise to a considerable administrative burden. A focused asset-sharing approach, whereby only vehicles are part of the sharing agreement, would effectively circumvent the issue at hand.

From today's perspective, it is uncertain how autonomous driving will advance and to what extent truck drivers can be replaced in the long term. If this is not the case, the labour law situation must be conceptually considered in a TS environment. Against this background, it is relevant to consider labour law conditions as soon as personnel such as truck drivers become part of a TS concept.

Implications and recommendations

The main contribution of TS lies in its potential to reduce truck downtime and provide greater flexibility in transport planning, thus enabling more efficient resource allocation.

Stakeholder-integrated coping with barriers: The analysis revealed the existence of functional and perceptual barriers relevant at the organisational and individual levels. Whereas functional barriers affect primarily the organisational level, perceptual barriers are relevant at the individual level, specifically for truck drivers. Overcoming these barriers is crucial for leveraging TS.

Application of the TAM highlighted that PU is a critical predictor of behavoiursl intention to use TS. Even if TS is objectively useful in terms of transport efficiency, it might not be adopted if users do not perceive it as useful. This distinction between actual and PU is crucial to derive TS strategies, given that overcoming both functional and perceptual barriers appears necessary for successful TS adoption. To address barriers effectively, strategies must be implemented at both the organisational and individual levels, by considering the specific needs of each stakeholder group. Whereas the decision to adopt TS is ultimately a business choice, its success depends on gaining truck drivers' acceptance by aligning the system's value with their needs. Thus, designing TS requires a comprehensive approach integrating both company objectives and truck drivers' concerns.

Conceptual implications: The findings underscore the essential role of a digital platform in facilitating TS. This platform is essential for coordinating and managing available freight vehicles, enabling efficient vehicle allocation, and ensuring compliance with transport requirements and restrictions by matching supply and demand in real time. Moreover, the platform supports communication among stakeholders, thereby promoting transparency and trust. Trust is critical to the success of TS, and includes trust in the condition of the vehicles, the reliability of the system, and other participants. Without this trust, commitment to TS would be limited, and potential benefits would be restricted. Conceptual platform recommendations focus on these aspects to ensure effective functioning of the system. The following solutions have shown to bridge the gap between the normative vision and the status quo within the industry:

1. *Keyless and contactless vehicle takeover*: App-based keyless handover significantly reduces administrative costs, thus making ultra-short-term truck rentals feasible.
2. *Image/video recording as documentary evidence*: To address concerns regarding vehicle cleanliness and damage, an app-integrated system allows truck drivers to document the vehicle's condition before and after use. AI-driven comparisons minimise disputes, thereby promoting accountability and ensuring that vehicles are returned in good condition.
3. *Processing of financial transactions*: Managing payments through the TS platform builds trust and simplifies transactions. The platform acts as a secure intermediary, thereby reducing risks and enhancing user experience.
4. *Integrated vehicle insurance solution:* An integrated vehicle insurance solution covers vehicles during sharing, and consequently minimises administrative burdens and risks for both owners and users. Full-coverage options with capped liability protect owners from unexpected costs and ensure a reliable sharing process.
5. *Fuel card-based refuelling process*: Platform-managed fuel cards ensure transparent fuel costs, integrated with rental payments. Telematics systems monitor fuel levels, with penalties for non-compliance, thus ensuring that vehicles are ready for immediate use and maintaining reliability in TS.
6. *Toll accounting and liability*: To manage toll payments, credit checks and contractual agreements are necessary. The platform ensures timely toll payments, thus reducing financial risks and securing trust within the TS system, particularly given the short rental periods involved.

5.2 Contribution of this Thesis

From a theoretical perspective, this thesis integrates the engaged scholars' approach with backcasting within a mixed-methods design. This combination has proven effective in exploring and establishing a new research field: truck-sharing in road freight transport. The methodology offered a pragmatic approach to address specific practical problems, rooted in entrepreneurial practice with a future-oriented perspective. Rather than pursuing a purely visionary concept, this thesis tackles persistent efficiency deficits in freight transport by building on established practices such as horizontal collaboration and expanding the sharing economy into new domains.

While the sharing economy has garnered significant attention from both scholars and practitioners, this thesis pioneers its application in road freight transport, establishing TS as a legitimate field of research. It provides a conceptual foundation for TS, addressing real-world challenges and offering a forward-looking solution to current inefficiencies in freight transport operations.

By identifying and addressing barriers to TS adoption, the thesis contributes to the development of practical solutions for future implementation. It projects conceptual use cases for TS, aiming to unlock its potential and offering insights into its gradual integration into business practices. While the primary focus is on business applications, the thesis also acknowledges the ecological and economic benefits of TS, which can be realised through future adoption in the industry.

5.3 Practical Implications

For practitioners, TS is a forward-looking concept with significant potential benefits for companies engaged in road freight transport. The research addresses persistent inefficiencies in truck utilisation, highlighting how TS can reduce vehicle downtime and increase flexibility in transport and capacity planning. It is possible for fleet operators to achieve higher rates of truck utilisation and to reduce the number of empty kilometres travelled, which will have a positive impact on both cost efficiency and environmental sustainability.

The research highlights the necessity of a strategic approach to fleet and capacity management. By examining TS through a practical lens, the thesis provides a conceptual foundation for companies to explore new approaches to operational optimisation. This includes the identification of specific, future-oriented use cases that can serve as models for the effective integration of TS into existing business practices. The insights gained here can assist LSPs in developing transport strategies that are responsive to evolving market demands.

Furthermore, the thesis emphasises the significant function of digital platforms in facilitating the practical implementation of TS. An integrated platform is identified as the foundation of the TS model, facilitating real-time coordination, compliance, and trust among stakeholders. For organisations seeking to implement TS, it is crucial to comprehend the technological infrastructure required and the ways in which it can be leveraged to optimise operational efficiency.

The thesis also addresses the human factor, particularly the roles and perceptions of truck drivers in the success of TS. In the context of an increasing truck driver shortage, the thesis emphasises the need for inclusive strategies that consider the drivers' perspectives and concerns. By doing so, companies can foster a more supportive environment for TS adoption, ensuring that this innovation is not only technically feasible but also socially sustainable.

In a broader perspective, the thesis encourages practitioners to view TS not just as a solution to operational inefficiencies but as part of a larger shift towards more collaborative and flexible business models in freight transport. By engaging with TS, companies can position themselves at the forefront of this shift. In conclusion, the practical implications of this thesis extend beyond immediate operational improvements. They offer a pragmatic pathway for companies to rethink their approach to freight transport, embracing new technologies and collaborative models that promise to redefine efficiency and sustainability in road freight transport.

5.4 Limitations

While this thesis is grounded in real-world problems, persistent efficiency deficits in freight transport, building on horizontal cooperation and the sharing economy as established practices, it is subject to limitations. The following paragraphs outline the content and methodology-related limitations concerning the research approach and findings.

This thesis adopts an economic perspective as its starting point and objective. It draws on efficiency deficits to examine economic parameters that are used as a benchmark, deriving the TS concept from this and identifying drivers and barriers. Beyond the directly addressed economic contribution, TS is assumed to have ecological benefits. However, the ecological dimension of efficiency was not the central focus of this thesis, nor was social efficiency. Therefore, the narrowing of the scope is a limitation that must be acknowledged.

Further limitations pertain to the methodology employed. Given that TS is an abstract phenomenon that cannot be empirically observed, backcasting, a future-oriented research approach, was utilised to investigate truck-sharing as a normative solution in the context of road freight transport. Firstly, it was not possible to evaluate the feasibility of implementing the proposed organisational and technological design for the TS system beyond expert interviews within the context of this thesis, as phenomenological existence was not observed in practice. Secondly, it is typical for the use of backcasting to result in a time lag of years or even decades with regard to implementation. Conversely, deriving insights and implications for prospective implementation is based on the perspective of the present and the prevailing conditions. The proposed implementation of TS is subject to the prevailing conditions and value patterns of the time. While backcasting methodology assumes future scenarios, prospective dynamics cannot be reliably predicted. Legislation and socially anchored values and beliefs may change over time, rendering the assumptions and identified measures obsolete due to unexpected external changes, thus rendering the practical solutions unfeasible.

Further limitations are evident in the composition of the interview partners and the sample of the online survey. Although the list of interviewees was carefully compiled to include all relevant stakeholder perspectives (e.g. LSPs, truck drivers, shippers, authorities), the interviewees were concentrated geographically in the German-speaking area. Therefore, the validity of the results is only substantiated in this area and is limited in terms of geographical scope.

The interviewing process offered profound insights into a complex and previously unexamined research phenomenon, which proved to be a valuable addition to the existing body of knowledge. However, this carries the risk of the findings being limited in their generalisability, given that the statements made by the interviewees are shaped by their individual experiences and value patterns. This shortcoming was addressed from a research design perspective by conducting a subsequent online survey to validate preliminary findings. However, it is important to note that the survey sample was not randomly selected but rather based on convenience. Self-selection may have resulted in the over- or underrepresentation of certain groups, and the potential for selection effects cannot be entirely discounted.

This thesis lays the groundwork for TS and sets the course for future research to build on and extend its contributions. It proposes the establishment of industry consortia with the objective of advancing the exploration and conceptualisation of TS practice.

5.5 Opportunities for Future Research

Based on the research findings and limitations of this thesis further research avenues open up. Future research may consider the following directions:

Differentiated and actor-centred use case exploration: In this thesis, conceptual use cases were developed with the aim of addressing existing and persisting inefficiencies in road freight transport. In light of the variety of different actors, business models and transport requirements, it can be assumed that the potentials and applications differ from company to company. Future research should aim for a differentiated analysis of fleet operator groups in order to explore use cases for each actor. For LSPs, providing transport services is their core business, whereas own-account transport companies often operate transport as a downstream function alongside their actual core business. A differentiated analysis of potential could show for which players TS is particularly suitable.

Economic impact and cost-benefit analyses: Another avenue for future investigation is to conduct comprehensive economic analyses with the objective of quantifying the financial benefits of truck sharing in general and for businesses of varying sizes and types. This could examine how TS can result in long-term savings and provide insights into the potential impact on overall profitability. By evaluating the economic viability of TS in different business models and industries, future studies can provide a clearer understanding of the potential of TS to improve operational efficiency and financial performance in the road freight sector. In this context, emerging cost drivers must also be taken into account, since the organisation of a TS system entails effort and costs for the operation of a platform.

Expansion of the efficiency analysis beyond the focused economic consideration: A sustainability-oriented strategy could adopt a holistic view of economic, ecological, and social targets. Analysing the ecological advantages of truck sharing, particularly in terms of reducing emissions and traffic, would appear to be a logical first step.

Sharing in road freight transport beyond trucks: As the engine transformation is likely to progress in the future, it may be reasonably assumed that the focus on sharing in road freight transport will be extended beyond tucks. Currently, the proportion of electric freight vehicles is relatively low in most European countries, though it is increasing. Reliable charging infrastructure constitutes a crucial factor in the adoption of electric freight vehicles, contingent on the assumption that a dense network of charging infrastructure is in place. In consequence, the question of sharing charging infrastructure remains a topic of discussion.

References

Acquier, A., Daudigeos, T., & Pinkse, J. (2017). Promises and paradoxes of the sharing economy: An organizing framework. *Technological Forecasting and Social Change, 125*, 1–10. https://doi.org/10.1016/j.techfore.2017.07.006

ADAC (2024, February 2). *ADAC Staubilanz 2023: Deutschland-Ticket reduziert Staus nicht.* https://www.adac.de/verkehr/verkehrsinformationen/staubilanz/

Agarwal, R., & Ergun, Ö. (2010). Network design and allocation mechanisms for carrier alliances in liner shipping. *Operations research, 58*(6), 1726–1742. https://doi.org/10.1287/opre.1100.0848

Alarcón, F. E., Mac Cawley, A., & Sauma, E. (2023). Electric mobility toward sustainable cities and road-freight logistics: A systematic review and future research directions. *Journal of Cleaner Production, 430*, 138959. https://doi.org/10.1016/j.jclepro.2023.138959

Aloui, A., Hamani, N., Derrouiche, R., & Delahoche, L. (2021). Systematic literature review on collaborative sustainable transportation: Overview, analysis and perspectives. *Transportation Research Interdisciplinary Perspectives, 9*, 100291. https://doi.org/10.1016/j.trip.2020.100291

Arbeitskammer Wien (2001). *Transportpreise und Transportkosten der verschiedenen Verkehrsträger im Güterverkehr*. Retrieved 25.06.2023 from https://docplayer.org/26981771-Transportpreise-und-transportkosten-der-verschiedenen-verkehrstraeger-im-gueterverkehr.html

Arvidsson, N., Woxenius, J., & Lammgård, C. (2013). Review of Road Hauliers' Measures for Increasing Transport Efficiency and Sustainability in Urban Freight Distribution. *Transport Reviews, 33*(1), 107–127. https://doi.org/10.1080/01441647.2013.763866

Avey, J. B., Avolio, B. J., Crossley, C. D., & Luthans, F. (2009). Psychological ownership: Theoretical extensions, measurement and relation to work outcomes. *Journal of Organizational Behavior, 30*(2), 173–191. https://doi.org/10.1002/job.583

Bachmann, R. (2010). Verdeckter Personalverleih: Aspekte zur rechtlichen Ausgestaltung, zur Bewilligungspflicht, zum Konzernverleih und zum Verleih mit Auslandsberührung. In J.-F. Stöckli (Ed.), *Mitteilungen des Instituts für Schweizerisches Arbeitsrecht* (pp. 53–97). Stämpfli Verlag.

Bagozzi, R. P., & Yi, Y. (1988). On the evaluation of structural equation models. *Journal of the academy of marketing science, 16*, 74–94. https://doi.org/10.1007/BF02723327

L. Häberle, *Truck-Sharing in Road Freight Transport*, Supply Chain Management, https://doi.org/10.1007/978-3-658-48754-6

Ballot, E., & Fontane, F. (2010). Reducing transportation CO2 emissions through pooling of supply networks: Perspectives from a case study in French retail chains. *Production Planning & Control, 21*, 640–650. https://doi.org/10.1080/09537287.2010.489276

Barrella, E., & Amekudzi, A. A. (2011). Backcasting for sustainable transportation planning. *Transportation Research Record, 2242*, 29–36. https://doi.org/10.3141/2242-04

Bartlett, C., & Ghoshal, S. (2004). Managing across Boundaries: the Collaborative Challenge. In C. Bartlett, S. Ghoshal & J. Birkinshaw (Eds.), *Text, Cases and Readings in Cross-Border Management* (pp. 403–502). McGraw-Hill.

Basso, F., D'Amours, S., Rönnqvist, M., & Weintraub, A. (2019). A survey on obstacles and difficulties of practical implementation of horizontal collaboration in logistics. *International Transactions in Operational Research, 26*(3), 775–793. https://doi.org/10.1111/itor.12577

BCG (2022, October 24). *Mapping the Future of Autonomous Trucking.* https://www.bcg.com/publications/2022/mapping-the-future-of-autonomous-trucks

Belk, R. (2014). Sharing versus pseudo-sharing in Web 2.0. *The anthropologist, 18*(1), 7–23. https://doi.org/10.1080/09720073.2014.11891518

Benoit, S., Baker, T., Bolton, R., Gruber, T., & Kandampully, J. (2017). A Triadic Framework for Collaborative Consumption—Motives, Roles and Resources. *Journal of Business Research, 79*(1), 219–227. https://doi.org/10.1016/j.jbusres.2017.05.004

Berger, S. (2009). *Kooperative Tourenplanung—Eine quantitative Analyse* (Dissertation University of Halle-Wittenberg). https://doi.org/10.25673/42

Bernerth, J. B., & Aguinis, H. (2016). A critical review and best-practice recommendations for control variable usage. *Personnel Psychology, 69*, 229–283. https://doi.org/10.1111/peps.12103

Börjeson, L., Höjer, M., Dreborg, K. H., Ekvall, T., & Finnveden, G. (2006). Scenario types and techniques: towards a user's guide. *Futures, 38*(7), 723–739. https://doi.org/10.1016/j.futures.2005.12.002

Brown, G., Crossley, C., & Robinson, S. L. (2014). Psychological ownership, territorial behavior, and being perceived as a team contributor: The critical role of trust in the work environment. *Personnel psychology, 67*(2), 463–485. https://doi.org/10.1111/peps.12048

Brown, G., Lawrence, T. B., & Robinson, S. L. (2005). Territoriality in organizations. *Academy of Management Review, 30*(3), 577–594. https://doi.org/10.5465/amr.2005.17293710

Bundesamt für Logistik und Mobilität [BALM]. (2024, January). *Auswertung der Arbeitsbedingungen in Güterverkehr und Logistik 2023—Fahrerberufe.* https://www.balm.bund.de/SharedDocs/Downloads/DE/Marktbeobachtung/Turnusberichte_Arbeitsbedingungen/AGL_2023.html?nn=3290874

Bundesamt für Statistik [BFS]. (2023a, November 27). *Güterverkehr in der Schweiz 2022.* https://www.bfs.admin.ch/bfs/de/home/statistiken/mobilitaet-verkehr/gueterverkehr/strasse/schwere-fahrzeuge.assetdetail.29365208.html

Bundesamt für Statistik [BFS]. (2023b, November 27). *Modal Split Strasse—Schiene (Leistungen der Güterfahrzeuge. Zeitreihen)—1985–2022.* Bundesamt für Statistik. https://www.bfs.admin.ch/bfs/de/home/statistiken/mobilitaet-verkehr/gueterverkehr.assetdetail.28005378.html

Bundesamt für Strassen [ASTRA]. (2024). *Bericht Verkehrsentwicklung und Verkehrsfluss 2023.* https://www.astra.admin.ch/astra/de/home/dokumentation/medienmitteilungen/anzeige-meldungen.msg-id-101429.html

Bundesamt für Raumentwicklung [ARE]. (2022, April 8). *Verkehrsperspektiven 2050.* https://www.are.admin.ch/are/de/home/mobilitaet/grundlagen-und-daten/verkehrsperspektiven.html

Bundesministerium für Digitales und Verkehr [BMDV]. (2023, March 1). *Gleitende Langfrist-Verkehrsprognose 2021–2022.* https://bmdv.bund.de/SharedDocs/DE/Anlage/K/prognose-berichtgleitende-langfrist-verkehrsprognose.pdf?__blob=publicationFile

Bundesverband Güterkraftverkehr Logistik und Entsorgung [BGL] (2024). *Branchenkostenentwicklung: Modellrechnungen zur Kostenentwicklung im Güterkraftverkehr.* Retrieved 25.06.2024 from https://www.bgl-ev.de/interaktiver-branchenkostenmodellrechner/

Caputo, J., & Mininno, M., (1996). Internal, vertical and horizontal logistics integration in Italian grocery distribution. *International Journal of Physical Distribution & Logistics Management, 26*, 64–90. https://doi.org/10.1108/09600039610149101

Carissimi, M. C., & Creazza, A. (2022). The role of the enabler in sharing economy service triads: A logistics perspective. *Cleaner logistics and supply chain, 5*, 100077. https://doi.org/10.1016/j.clscn.2022.100077

Chabot, T., Bouchard, F., Legault-Michaud, A., Renaud, J., & Coelho, L. C. (2018). Service level, cost and environmental optimization of collaborative transportation. *Transportation Research Part E: Logistics and Transportation Review, 110*, 1–14. https://doi.org/10.1016/j.tre.2017.11.008

Christopher, M. (2011). *Logistics and Supply Chain Management* (4th ed.). Pearson Education.

Cohen, J. (1988). Statistical power analysis for the behavioral Sciences (2nd ed). Lawrence Erlbaum Associates.

Creswell, J., (2007) *Qualitative Inquiry and Research Design: Choosing among five approaches.* (2nd ed.) SAGE Publications.

Creswell, J. W., & Clark, V. L. P. (2017). Designing and conducting mixed methods research (3rd ed.). SAGE Publications.

Cruijssen, F. (2006). *Horizontal cooperation in transport and logistics* (Dissertation Tilburg University). Center for Economic Research. https://research.tilburguniversity.edu/en/publications/horizontal-cooperation-in-transport-and-logistics

Cruijssen, F., Bräysy, O., Dullaert, W., Fleuren, H., & Salomon, M. (2007). Joint route planning under varying market conditions. *International Journal of Physical Distribution and Logistics Management, 37*(4), 287–304. https://doi.org/10.1108/09600030710752514

Cruijssen, F., Cools, M., & Dullaert, W. (2007). Horizontal Cooperation in Logistics: Opportunities and Impediments. *Transportation Research Part E Logistics and Transportation Review, 43*(2), 129–142. https://doi.org/10.1016/j.tre.2005.09.007

Cruijssen, F., Dullaert, W., & Fleuren, H. (2007). Horizontal Cooperation in Transport and Logistics: A Literature Review. *Transportation Journal, 46*(3), 22–39. https://doi.org/10.2307/20713677

Cruijssen, F., Dullaert, W., & Joro, T. (2010). Freight transportation efficiency through horizontal cooperation in Flanders. *International Journal of Logistics Research and Applications, 13*(3), 161–178. https://doi.org/10.1080/13675560903224962

Daher, S., Hausmann, L., & Wölfel, T. (2022, August 31). *Freight forwarders' earnings amid carrier-rate volatility.* McKinsey. https://www.mckinsey.com/industries/travel-logistics-and-infrastructure/our-insights/freight-forwarders-earnings-amid-carrier-rate-volatility/

Dai, B., & Chen, H. (2012). Proft allocation mechanisms for carrier collaboration in pickup and delivery service. *Computers & Industrial Engineering, 62*, 633–643. https://doi.org/10.1016/j.cie.2011.11.029

Daudi, M. (2018). *Trust in sharing resources in logistics collaboration* (Dissertation, University of Bremen). http://nbn-resolving.de/urn:nbn:de:gbv:46-00106594-13

Davis, F. D. (1989). Perceived usefulness, perceived ease of use, and user acceptance of information technology. *MIS quarterly, 13*(3), 319–340. https://doi.org/10.2307/249008

Davis, F. D., & Venkatesh, V. (1996). A critical assessment of potential measurement biases in the technology acceptance model: three experiments. *International journal of human-computer studies, 45*(1), 19–45. https://doi.org/10.1006/ijhc.1996.0040

Dawkins, S., Tian, A. W., Newman, A., & Martin, A. (2015). Psychological ownership: A review and research agenda. *Journal of Organizational Behavior, 38*(2), 163–183. https://doi.org/10.1002/job.2057

Deepen, J.M., Goldsby, T.J., Knemeyer, A.M., & Wallenburg, C.M. (2008). Beyond expectations: an examination of logistics outsourcing goal achievement and goal exceedance. *Journal of Business Logistics, 29*(2), 75–105. https://doi.org/10.1002/j.2158-1592.2008.tb00088.x

Defee, C., Williams, B., Randall, W.S., & Thomas, R. (2010). An inventory of theory in logistics and SCM research. *The International Journal of Logistics Management, 21*(3), 404–489. https://doi.org/10.1108/09574091011089817

Destatis (2024). *Lkw-Maut-Fahrleistungsindex.* Retreived 01.07.2024 from https://www.destatis.de/DE/Service/EXSTAT/Datensaetze/lkw-maut-artikel.html

Deutsches Verkehrsforum (2024, July 3). *Weckruf der Verbände: Bundeshaushalt 2025 muss der verkehrlichen Realität standhalten.* https://www.verkehrsforum.de/de/service/positionen/2024-07-03-weckruf-der-verbaende

Dong, C., Akram, A., Andersson, D., Arnäs, P. O., & Stefansson, G. (2021). The impact of emerging and disruptive technologies on freight transportation in the digital era: current state and future trends. *The International Journal of Logistics Management, 32*(2), 386–412. https://doi.org/10.1108/IJLM-01-2020-0043

Dossabhoy, N. S., & Berger, P. D. (2002). Business school research: bridging the gap between producers and consumers. *Omega, 30*(4), 301–314. https://doi.org/10.1016/s0305-0483(02)00031-2

Ergun, O., Kuyzu, G., & Savelsbergh, M. (2007). Reducing truckload transportation costs through collaboration. *Transportation science, 41*(2), 206–221. https://doi.org/10.1287/trsc.1060.0169

European Commission (2011, March 28). *Roadmap to a Single European Transport Area—Towards a Competitive and Resource Efficient Transport System.* Brussels: European Commission. https://eur-lex.europa.eu/legal-content/EN/TXT/?uri=CELEX%3A52011DC0144

European Commission (2023). *EU transport in figures: statistical pocketbook 2023.* Publications Office of the European Union. https://data.europa.eu/doi/10.2832/319371

European Environment Agency [EEA]. (2010, October 4). *Load factors for freight transport.* http://www.eea.europa.eu/data-and-maps/indicators/load-factors-for-freight-transport/load-factorsfor-freight-transport-1

Eurostat (2023, October). *Road freight transport by journey characteristics.* https://ec.europa.eu/eurostat/statistics-explained/index.php?title=Road_freight_transport_by_journey_characteristics#Empty_runnings_of_road_freight_vehicles

Faul, F., Erdfelder, E., Lang, A. G., & Buchner, A. (2007). G* Power 3: A flexible statistical power analysis program for the social, behavioral, and biomedical sciences. *Behavior research methods, 39*, 175–191. https://doi.org/10.3758/BF03193146

Fishbein, M., & Ajzen, I. (1975). *Belief, Attitude, Intention, and Behavior: An Introduction to Theory and Research.* Addison-Wesley.

Freichel, S. (1992). *Organisation von Logistikservice-Netzwerken: Theoretische Konzeption und empirische Fallstudien.* Erich Schmidt Verlag.

Freitag, M., Kück, M., & Becker, T. (2016). Potentials and Risks of Resource Sharing in Production and Logistics. In W. Delfmann & T. Wimmer (Eds.), *Proceedings of the 8th International Scientific Symposium on Logistics* (pp. 199–209). Bundesvereinigung Logistik.

Friedli, T., Zacharias, L., & Häberle, L. (2024, July 1). *Logistikmarktstudie Schweiz 02–2024: Trends, Nachhaltigkeit, Geopolitische Krisen, Automatisierung.* GS1 Switzerland. https://lms.gs1.ch/de

Gansterer, M., & Hartl, R. F. (2018). Collaborative vehicle routing: A survey. *European Journal of Operational Research, 268*, 1–12. https://doi.org/10.1016/j.ejor.2017.10.023

Gibson, B. J., Rutner, S.M., & Keller, S.B. (2002). Shipper-carrier partnership issues, rankings and satisfaction. *International Journal of Physical Distribution & Logistics Management, 32*(8), 669–681. https://doi.org/10.1108/09600030210444917

Gioia, D. A., Corley, K. G. & Hamilton, A. L. (2013). Seeking Qualitative Rigor in Inductive Research. *Organizational Research Methods, 16*(1), 15–31. https://doi.org/10.1177/1094428112452151

Gnamm, J., Stricker, K., & Bendig, O. (2010). *Zukunftsmarkt Lkw-Vermietung.* Bain & Company. https://www.bain.com/contentassets/50587e5ea5554d3d849de941f1263519/bain_brief_zukunftsmarktlkw-vermietung_2010_final.pdf

Graf, L., & Anner, F. (2021). Autonomous vehicles as the ultimate efficiency driver in logistics. In C. Wurst & L. Graf (Eds.), *Disrupting Logistics: Startups, Technologies, and Investors Building Future Supply Chains* (pp. 191–206). https://doi.org/10.1007/978-3-030-61093-7

Gray, D. E. (2013) *Doing Research in the Real World.* SAGE Publications.

Greening, P., Palmer, A., Piecyk, M., & McKinnon, A. C. (2015). An assessment of the potential for demand-side fuel savings in the Heavy Goods Vehicle (HGV) sector. *Centre for Sustainable Road Freight.* https://www.csrf.ac.uk/outputs/an-assessment-of-the-potential-for-demand-side-fuel-savings-in-the-heavy-goods-hgv-sector/

Greub, U. (2018). Personalverleih. In: W. Portmann & A. von Kaenel (Eds.), *Fachhandbuch Arbeitsrecht Expertenwissen für die Praxis* (pp. 607–634). Schulthess Juristische Medien AG.

Häberle, L., & Stölzle, W. (2024). Wettbewerb. In W. Stölzle, D. Schöder & L. Häberle (Eds.), *Güterverkehr kompakt* (pp. 191–203). De Gruyter Oldenbourg.

Hamari, J., Sjöklint, M., & Ukkonen, A. (2016). The sharing economy: Why people participate in collaborative consumption. *Journal of the Association for Information Science and Technology*, 67(9), 2047–2059 https://doi.org/10.1002/asi.23552

Herrmann, A., Brenner, W., & Stadler, R. (2018). *Autonomous driving: how the driverless revolution will change the world*. Emerald Publishing.

Herrmann, A., Jungwirth, J., & Huber, F. (2022). *Mobilität für alle:... auf Knopfdruck*. Campus Verlag.

Holmberg, J., & Robert, K. H. (2000). Backcasting—a framework for strategic planning. *International Journal of Sustainable Development and World Ecology, 7*(4), 291–308. https://doi.org/10.1080/13504500009470049

Hossain, M. (2020). Sharing economy: A comprehensive literature review. *International Journal of Hospitality Management, 87*, 102470. https://doi.org/10.1016/j.ijhm.2020.102470

International Energy Agency [IEA]. (2024, April). *Global EV Outlook 2024*. https://www.iea.org/reports/global-ev-outlook-2024

International Road Transport Union [IRU]. (2006, February). *Selected recent statistics on road freight transport in Europe*. https://www.iru.org/sites/default/files/2016-01/en-statistics-goods_0.pdf?fbclid=IwAR0ZTl7rM97s6s7of__sietJ5YWs4kosujfVt7QPgsllEqOBCSLgPfzgEE8

International Road Transport Union [IRU]. (2024, April 09). *Europe Truck Driver Shortage Report 2023*. https://www.iru.org/resources/iru-library/europe-truck-driver-shortage-report-2023

Islam, S. (2018). Simulation of truck arrival process at a seaport: evaluating truck-sharing benefits for empty trips reduction. *International Journal of Logistics Research and Applications, 21*(1), 94–112. https://doi.org/10.1080/13675567.2017.1353067

Islam, S., & Olsen, T. (2014). Truck-sharing challenges for hinterland trucking companies: A case of the empty container truck trips problem. *Business Process Management Journal, 20*(2), 290–334. https://doi.org/10.1108/BPMJ-03-2013-0042

Islam, S., Olsen, T., & Daud Ahmed, M. (2013). Reengineering the seaport container truck hauling process: Reducing empty slot trips for transport capacity improvement. *Business process management journal, 19*(5), 752–782. https://doi.org/10.1108/BPMJ-Jun-2012-0059

Islam, S., Shi, Y., Ahmed, J. U., & Uddin, M. J. (2019). Minimization of empty container truck trips: insights into truck-sharing constraints. *International Journal of Logistics Management, 30*(2), 641–662. https://doi.org/10.1108/IJLM-08-2018-0191

Islam, S., Uddin, M. J., Shi, Y., Sharif, T., & Ahmed, J. U. (2021). Exploring shippers' motivations to adopt collaborative truck-sharing initiatives. *International Journal of Physical Distribution & Logistics Management, 51*(5), 528–550. https://doi.org/10.1108/IJPDLM-10-2019-0303

Kampf, R., Potkány, M., Krajčírová, L., & Marcineková, K. (2016). Life cycle cost calculation and its importance in vehicle acquisition process for truck transport. *Naše More, 63*(3), 129–133. https://doi.org/10.17818/nm/2016/si10

Karam, A., Reinau, K. H., & Østergaard, C. R. (2021). Horizontal collaboration in the freight transport sector: barrier and decision-making frameworks. *European Transport Research Review, 13*(1), 1–22. https://doi.org/10.1186/s12544-021-00512-3

Kille, C., Schmidt, T., Stölzle, W., Häberle, L., & Rank, S. (2023). *Begegnung von Kapazitätsengpässen im Straßengüterverkehr—Fokus Personal.* Cuvillier.

King, N. (2004). Using Interviews in Qualitative Research. In C. Cassell & G. Symon (Eds.), *Essential guide to qualitative methods in organizational research* (pp. 11–22). SAGE Publications.

Kraftfahrt-Bundesamt [KBA]. (2023). *Inlandsverkehr Jahr 2022.* https://www.kba.de/DE/Statistik/Produktkatalog/produkte/Kraftverkehr/vd3_uebersicht.html

Kraftfahrt-Bundesamt [KBA]. (2024). *Bestand an Kraftfahrzeugen nach Umwelt-Merkmalen, 1. Januar 2024 (FZ 13).* https://www.kba.de/DE/Statistik/Fahrzeuge/Bestand/Umwelt/umwelt_node.htm

Lam, J. S. L., & Dai, J. (2015). Environmental sustainability of logistics service provider: an ANP-QFD approach. *The International Journal of Logistics Management, 26*(2), 313–333. https://doi.org/10.1108/ijlm-08-2013-0088

Lambert, D. M., Emmelhainz, M. A., & Gardner, J. T. (1999). Building successful logistics partnerships. *Journal of business logistics, 20*(1), 165–181. https://www.proquest.com/scholarly-journals/building-successful-logistics-partnerships/docview/212654249/se-2?accountid=28962

Landwehr, S., & Loew, T. (2024, August 6). *Die Qual der Fahrzeugwahl.* Deutsche Verkehrs-Zeitung. https://www.dvz.de/unternehmen/strasse/detail/news/die-qual-der-fahrzeugwahl.html

Lehoux, N., D'Amours, S., & Langevin, A. (2014). Inter-firm collaborations and supply chain coordination: Review of key elements and case study. *Production Planning & Control, 25*, 858–872. https://doi.org/10.1080/09537287.2013.771413

Lewin, K. (1947). Frontiers in group dynamics: Concept, method and reality in social science; social equilibria and social change. *Human relations, 1*(1), 5–41. https://doi.org/10.1177/001872674700100103

Liu, P., Wu, Y., & Xu, N. (2010). Allocating collaborative profit in less-than-truckload carrier alliance. *Journal of Service science and Management, 3*(1), 143–149.

Liu, Y., & Yang, Y. (2018). Empirical examination of users' adoption of the sharing economy in China using an expanded technology acceptance model. *Sustainability, 10*(4), 1262. https://doi.org/10.3390/su10041262

Looser, F. (2015). *Der Personalverleih: unter besonderer Berücksichtigung des GAV Personalverleih.* Helbling Lichtenhahn.

Lord, S., Després, C., & Ramadier, T. (2011). When mobility makes sense: A qualitative and longitudinal study of the daily mobility of the elderly. *Journal of Environmental psychology, 31*(1), 52–61. https://doi.org/10.1016/j.jenvp.2010.02.007

Makaci, M., Reaidy, P., Evrard-Samuel, K., Botta-Genoulaz, V., & Monteiro, T. (2017). Pooled warehouse management: An empirical study. *Computers & Industrial Engineering, 112*, 526–536. https://doi.org/10.1016/j.cie.2017.03.005

Marangunić, N., & Granić, A. (2015). Technology acceptance model: a literature review from 1986 to 2013. *Universal access in the information society, 14*, 81–95. https://doi.org/10.1007/s10209-014-0348-1

Mathiassen, L. (2017). Designing engaged scholarship: From real-world problems to research publications. *Engaged Management Review, 1*(1), 17–28. https://doi.org/10.28953/2375-8643.1000

Matusiewicz, M. & Ksiazkiewicz, D. (2023). *Shared Logistics—Literature Review. Applied Sciences, 13*(4), 2036. https://doi.org/10.3390/app13042036

Mayring, P. (2000). Qualitative Content Analysis. *Forum Qualitative Sozialforschung Forum: Qualitative Social Research, 1*(2). https://doi.org/10.17169/fqs-1.2.1089

McKinnon, A. (2000). Sustainable distribution: opportunities to improve vehicle loading. *Industry and environment, 23*(4), 26–30.

McKinnon, A. (2018). *Decarbonizing logistics: Distributing goods in a low carbon world.* Kogan Page Publishers.

McKinnon, A. C., & Ge, Y. (2004). Use of a synchronised vehicle audit to determine opportunities for improving transport efficiency in a supply chain. *International Journal of Logistics Research and Applications, 7*(3), 219–238. https://doi.org/10.1080/13675560412331298473

Melander, L., & Wallström, H. (2023). The benefits of green horizontal networks: Lessons learned from sharing charging infrastructure for electric freight vehicles. *Business Strategy and the Environment, 32*(4), 1835–1846. https://doi.org/10.1002/bse.3222

Mendoza, J. M. F., Sharmina, M., Gallego-Schmid, A., Heyes, G., & Azapagic, A. (2017). Integrating backcasting and eco-design for the circular economy: The BECE framework. *Journal of Industrial Ecology, 21*(3), 526–544. https://doi.org/10.1111/jiec.12590

Mentzer, J. T., Foggin, J. H., & Golicic, S. L. (2000). Collaboration: the enablers, impediments, and benefits. *Supply chain management review, 4*(4), 52–58.

Miles, M. B., & Huberman, A. M. (1994). *Qualitative data analysis: An expanded sourcebook* (2nd ed.). SAGE Publications.

Mirzabeiki, V., & Aitken, J. (2023). Panarchy-based transformative supply chain resilience: the role of supply chain capital. *International Journal of Operations & Production Management, 43*(1), 99–139. https://doi.org/10.1108/IJOPM-03-2022-0191

Montoya-Torres, J. R., Muñoz-Villamizar, A., & Vega-Mejía, C. A. (2016). On the impact of collaborative strategies for goods delivery in city logistics. *Production Planning & Control, 27*, 443–455. https://doi.org/10.1080/09537287.2016.1147092

Mrabti, N., Hamani, N., & Delahoche, L. (2022). A sustainable collaborative approach to the distribution network design problem with CO2 emissions allocation. *International journal of shipping and transport logistics, 14*(1–2), 114–140. https://doi.org/10.1504/IJSTL.2022.120676

Muñoz, P., & Cohen, B. (2017). Mapping out the sharing economy: A configurational approach to sharing business modeling. *Technological forecasting and social change, 125*, 21–37. https://doi.org/10.1016/j.techfore.2017.03.035

Musso, A. (2001). SOFTICE: survey on freight transport including a cost comparison for Europe. *Transportation research record, 1763*(1), 27–34. https://doi.org/10.3141/1763-05

Naesens, K., Gelders, L., & Pintelon, L. (2009). A swift response framework for measuring the strategic fit for a horizontal collaborative initiative. *International Journal of Production Economics, 121*(2), 550–561. https://doi.org/10.1016/j.ijpe.2007.04.004

Nansubuga, B., & Kowalkowski, C. (2021). Carsharing: a systematic literature review and research agenda. *Journal of Service Management, 32*(6), 55–91. https://doi.org/10.1108/JOSM-10-2020-0344

Ocicka, B., & Wieteska, G. (2017). Sharing economy in logistics and supply chain management. *Logforum, 13*(2), 183–193. https://doi.org/10.17270/J.LOG.2017.2.6

Oliver, R. K, & Webber, M. D. (1982). Supply chain management: logistics catches up with strategy. In M. Christopher (Ed.), *Logistics: The Strategic Issues* (pp. 63–75). Chapman & Hall.

Pahlén, P.-O., & Börjesson, F. (2012). Measuring resource efficiency in long haul road freight transport. *Proceedings of the 24th Annual Nordic Logistics Research Network (NOFOMA) Conference*. Naantali, Finland.

Pan, S., Trentesaux, D., Ballot, E., & Huang, G. Q. (2019). Horizontal collaborative transport: Survey of solutions and practical implementation issues. *International Journal of Production Research, 57*, 5340–5361. https://doi.org/10.1080/00207543.2019.1574040

Parkhe, A. (1993). Strategic alliance structuring: A game theoretic and transaction cost examination of interfirm cooperation. *Academy of management journal, 36*(4), 794–829. https://doi.org/10.2307/256759

Paulsen, N. (2019, April 16). *Expeditors und DSV bleiben die Margen-Könige*. Deutsche Verkehrs-Zeitung. https://www.dvz.de/unternehmen/detail/news/expeditors-und-dsv-bleiben-die-margen-koenige.html

Paundra, J., Rook, L., van Dalen, J., & Ketter, W. (2017). Preferences for car sharing services: Effects of instrumental attributes and psychological ownership. *Journal of environmental psychology, 53*, 121–130. https://doi.org/10.1016/j.jenvp.2017.07.003

Pfohl, H.-C. (Ed.). (2003). *Güterverkehr—Eine Integrationsaufgabe für die Logistik*. Erich Schmidt Verlag.

Pfohl, H.-C. (2018). *Logistiksysteme* (9th ed.). Springer Vieweg.

Piecyk, M. I. (2010) *Analysis of long-term freight transport, logistics and related CO2 trends on a business-as-usual basis* (Dissertation Heriot-Watt University). https://www.ros.hw.ac.uk/bitstream/handle/10399/2326/PiecykMI_0510_sml.pdf?sequence=1&isAllowed=y

Pierce, J. L., Kostova, T., & Dirks, K. T. (2003). The state of psychological ownership: Integrating and extending a century of research. *Review of general psychology, 7*(1), 84–107. https://doi.org/10.1037/1089-2680.7.1.84

Pomponi, F., Fratocchi, L., & Tafuri, S. R. (2015). Trust development and horizontal collaboration in logistics: A theory based evolutionary framework. *Supply Chain Management, 20*(1), 83–97. https://doi.org/10.1108/SCM-02-2014-0078

Qasim, M., & Csiszar, C., 2021. Major barriers in adoption of electric trucks in logistics system. *Promet—Traffic&Transportation, 33*(6), 833–846. https://doi.org/10.7307/ptt.v33i6.3922

Reimann, S. (2022, October 17). *Logistik-Margen: Konzerne und Kleine liegen vorne*. Deutsche Verkehrs-Zeitung. https://www.dvz.de/unternehmen/logistik/detail/news/logistik-margen-konzerne-und-kleine-liegen-vorne.html

Renko, S. (2011). Vertical Collaboration in the Supply Chain. In S. Renko (Ed.), *Supply Chain Management—New Perspectives* (pp. 183–198). InTech Open. https://doi.org/10.5772/20866

Robinson, J. B. (1990). Futures under glass. A recipe for people who hate to predict. *Futures, 22*(8), 820–842. https://doi.org/10.1016/0016-3287(90)90018-D

Robinson, J. (2003). Future subjunctive: backcasting as social learning. *Futures, 35*(8), 839–856. https://doi.org/10.1016/s0016-3287(03)00039-9

Robson, C. (2002). *Real World Research: A Resource for Social Scientists and Practitioner-Researchers* (2nd ed.). Blackwell Publishing.

SAE International (2021, May 3). *SAE Levels of Driving Automation™ Refined for Clarity and International Audience.* https://www.sae.org/blog/sae-j3016-update

Saunders, M., Lewis, P., & Thornhill, A. (2023). *Research Methods for Business Students* (9th ed.). Pearson.

Schmoltzi, C., & Wallenburg, C. M. (2011). Horizontal cooperations between logistics service providers: motives, structure, performance. *International Journal of Physical Distribution & Logistics Management, 41*(6), 552–575. https://doi.org/10.1108/09600031111147817

Schmoltzi, C., & Wallenburg, C. M. (2012). Operational governance in horizontal cooperations of logistics service providers: performance effects and the moderating role of cooperation complexity. *Journal of Supply Chain Management, 48*(2), 53–74. https://doi.org/10.1111/j.1745-493X.2011.03262.x

Schneider, C. (2024). Organisation. In W. Stölzle, D. Schöder & L. Häberle (Eds.), *Güterverkehr kompakt* (pp.69–86). De Gruyter Oldenbourg.

Schuster, A. M., Agrawal, S., Britt, N., Sperry, D., Van Fossen, J. A., Wang, S., Mack, E. A., Libermann, J., & Cotten, S. R. (2023). Will automated vehicles solve the truck driver shortages? Perspectives from the trucking industry. *Technology in Society, 74*, 102313. https://doi.org/10.1016/j.techsoc.2023.102313

Schweizerischer Bundesrat. (1985). *Botschaft zu einem revidierten Bundesgesetz über die Arbeitsvermittlung und den Personalverleih.* Fedlex. https://www.fedlex.admin.ch/eli/fga/1985/3_556_524_503/de

Schwemmer, M. (2018). *Die TOP 100 der Logistik 2018/2019—Marktgrößen, Marktsegmente und Marktführer.* DVV Media Group.

Simpson, J. R., Mishra, S., Talebian, A., & Golias, M. M. (2019). An estimation of the future adoption rate of autonomous trucks by freight organizations. *Research in Transportation Economics, 76,* 100737. https://doi.org/10.1016/j.retrec.2019.100737

Singh, H., Garg, R. K., & Sachdeva, A. (2018). Supply chain collaboration: A state-of-the-art literature review. *Uncertainity Supply Chain Management, 6*, 149–180. https://doi.org/10.5267/j.uscm.2017.8.002

Slowik, P., & Sharpe, B. (2018, March 28). *Automation in the long haul: Challenges and opportunities of autonomous heavy duty trucking in the United States.* ICCT. https://theicct.org/publication/automation-in-the-long-haul-challenges-and-opportunities-of-autonomous-heavy-duty-trucking-in-the-united-states/

Srinivas, S., Ramachandiran, S., & Rajendran, S. (2022). Autonomous robot-driven deliveries: A review of recent developments and future directions. *Transportation research part E: logistics and transportation review, 165*, 102834. https://doi.org/10.1016/j.tre.2022.102834

Staatsekretariat für Wirtschaft [SECO]. (2003). *SECO Weisungen und Erläuterungen zum Arbeitsvermittlungsgesetz, zur Arbeitsvermittlungsverordnung und der Gebührenverordnung zum Arbeitsvermittlungsgesetz.* https://www.seco.admin.ch/seco/de/home/Publikationen_Dienstleistungen/Publikationen_und_Formulare/Arbeit/Personenfreizuegigkeit_und_Arbeitsbeziehungen/merkblaetter/Weisungen_Erlaeuterungen_Arbeitsvermittlungsgesetz.html

Staatsekretariat für Wirtschaft [SECO] (2017). *Merkblatt zu den Bestimmungen über den Personalverleih gemäss Bundesgesetz über die Arbeitsvermittlung und den Personalverleih*

(AVG) vom 6. Oktober 1989. https://www.arbeit.swiss/secoalv/de/home/menue/arbeitsvermittler/private-arbeitsvermittlung-und-personalverleih.html

Stefansson, G. (2006). Collaborative logistics management and the role of third-party service providers. *International Journal of Physical Distribution & Logistics Management, 36*(2), 76–92. https://doi.org/10.1108/09600030610656413

Sternberg, H., Linan, I., Prockl, G., & Norrman, A. (2021). Tragedy of the facilitated commons: A multiple-case study of failure in systematic horizontal logistics collaboration. *Journal of Supply Chain Management, 57*(4), 3–22. https://doi.org/10.1111/jscm.12278

Stölzle, W., & Häberle, L. (2024). Kapazitätsmanagement. In W. Stölzle, D. Schöder & L. Häberle (Eds.), *Güterverkehr kompakt* (pp. 304–322). De Gruyter Oldenbourg.

Stölzle, W., Häberle, L., & Schöder, D. (2024). Handlungsoptionen und -restriktionen für Güterverkehrsunternehmen. In W. Stölzle, D. Schöder & L. Häberle (Eds.), *Güterverkehr kompakt* (pp. 45–52). De Gruyter Oldenbourg.

Stölzle, W., & Wildhaber, V. (2019). Premises for Truck Sharing in General Cargo Cooperatives–an Exploratory Case Study Research. In U. Clausen, S. Langkau & F. Kreuz (Eds.), *Advances in Production, Logistics and Traffic: Proceedings of the 4th Interdisciplinary Conference on Production Logistics and Traffic 2019 4* (pp. 75–89). Springer International Publishing.

Touboulic, A., McCarthy, L., & Matthews, L. (2020). Re-imaging supply chain challenges through critical engaged research. *Journal of Supply Chain Management, 56*(2), 36–51. https://doi.org/10.1111/jscm.12226

Ulrich, H. (1984). Die Betriebswirtschaftslehre als anwendungsorientierte Sozialwissenschaft. In T. Dyllick & G. Probst (Eds.), *Management* (pp. 168–199). Haupt.

Van de Ven, A. H. (2007). *Engaged scholarship: A guide for organizational and social research.* Oxford University Press.

Van de Ven, A. H. (2018). Academic-practitioner engaged scholarship. *Information and Organization, 28*(1), 37–43. https://doi.org/10.1016/j.infoandorg.2018.02.002

Van Hoek, R. (2021). Retrospective on the launch of IJPDLM–lessons for the future of logistics and supply chain management research. *International Journal of Physical Distribution & Logistics Management, 51*(10), 1065–1089. https://doi.org/10.1108/IJPDLM-11-2020-0349

Vendela, S. (2017) Towards more efficient logistics: increasing load factor in a shipper's road transport. *The International Journal of Logistics Management 28*(2), 228–250. https://doi.org/10.1108/IJLM-04-2015-0071

Venkatesh, V., & Davis, F. D. (2000). A theoretical extension of the technology acceptance model: Four longitudinal field studies. *Management science, 46*(2), 186–204. https://doi.org/10.1287/mnsc.46.2.186.11926

Verdonck, L., Caris, A., Ramaekers, K., & Janssens, G. K. (2013). Collaborative Logistics from the Perspective of Road Transportation Companies. *Transport Reviews, 33*(6), 700–719. https://doi.org/10.1080/01441647.2013.853706

Vergragt, P. J., & Quist, J. (2011). Backcasting for sustainability: Introduction to the special issue. *Technological forecasting and social change, 78*(5), 747–755. https://doi.org/10.1016/j.techfore.2011.03.010

Verstrepen, S., Cools, M., Cruijssen, F., & Dullaert, W. (2009). A dynamic framework for managing horizontal cooperation in logistics. *International Journal of Logistics Systems and Management, 5*(3–4), 228–248. https://doi.org/10.1504/IJLSM.2009.022497

Wang, X., & Kopfer, H. (2014). Collaborative transportation planning of less-than-truckload freight: A route-based request exchange mechanism. *OR spectrum, 36*, 357–380. https://doi.org/10.1007/s00291-013-0331-x

Wang, Y., Sanchez Rodrigues, V., & Evans, L. (2015). The use of ICT in road freight transport for CO2 reduction–an exploratory study of UK's grocery retail industry. *The International Journal of Logistics Management, 26*(1), 2–29. https://doi.org/10.1108/IJLM-02-2013-0021

Werwitzke, C. (2024, February 08). *Zulassungen von E-Nutzfahrzeugen: Das Europa der zwei Geschwindigkeiten.* https://www.electrive.net/2024/02/08/zulassungen-von-e-nutzfahrzeugen-das-europa-der-zwei-geschwindigkeiten/#:~:text=Bei%20den%20mittelschweren%20Lkw%20verzeichnete,der%20Neuzulassungen%20auf%2067.596%20Einheiten

Wilpert, B. (1991). Property, ownership, and participation: On the growing contradictions between legal and psychological ownership concepts. In R. Russell & V. Rus (Eds.), *International handbook of participations in organizations: For the study of organizational democracy, co-operation, and self management* (2nd ed., pp. 149–164). Oxford University Press.

Wohlgemuth, O. (2002). *Management netzwerkartiger Kooperationen—Instrumente für die unternehmensübergreifende Steuerung.* Gabler.

Wong, C. Y. (2021). Celebrating IJPDLM's 50th anniversary: a reflection on its contributions and future directions. *International Journal of Physical Distribution & Logistics Management, 51*(10), 1049–1064. https://doi.org/10.1108/IJPDLM-10-2021-0427

World Economic Forum (2021, April 1). Autonomous Trucks: An Opportunity to Make Road Freight Safer, Cleaner and More Efficient. https://www.weforum.org/publications/autonomous-trucks-an-opportunity-to-make-road-freight-safer-cleaner-and-more-efficient/

Wrenn, C. A. (2017). Can autonomous technology reduce the driver shortage in the commercial trucking industry (Dissertation California Southern University). https://www.researchgate.net/publication/319524313_Can_Autonomous_Technology_Reduce_the_Driver_Shortage_in_the_Commercial_Trucking_Industry

Yin, R. K. (2009). *Case study research: Design and methods* (5th ed.). SAGE Publications.

Yousafzai, S. Y., Foxall, G. R., & Pallister, J. G. (2007). Technology acceptance: a meta-analysis of the TAM: Part 2. *Journal of Modelling in Management, 2*(3), 281–304. https://doi.org/10.1108/17465660710834453

Zhang, Y., Liu, G., Zhang, L., Xu, S., & Cheung, M. W.-L. (2021). Psychological Ownership: A Meta-Analysis and Comparison of Multiple Forms of Attachment in the Workplace. *Journal of Management, 47*(3), 745–770. https://doi.org/10.1177/0149206320917195

Zhao, X., Ke, Y., Zuo, J., Xiong, W., & Wu, P. (2020). Evaluation of sustainable transport research in 2000–2019. *Journal of cleaner production, 256*, 120404. https://doi.org/10.1016/j.jclepro.2020.120404

Zineldin, M. (2004). Co-opetition: the organisation of the future. *Marketing intelligence & planning, 22*(7), 780–790. https://doi.org/10.1108/02634500410568600

The manufacturer's authorised representative in the EU is Springer
Nature Customer Service Centre GmbH, Europaplatz 3, 69115 Heidelberg,
Germany. If you have any concerns regarding our products, please
contact ProductSafety@springernature.com

Printed and bound by CPI Group (UK) Ltd, Croydon, CR0 4YY
01/10/2026
02225981-0002